The Building of a Dream
Journal of the Molly B
Volume IV

By
Tom Schmidt

Copyright © 2014

Frostproof, FL

ISBN 978-1-312-31443-6

All rights reserved.
No part of this material may be reproduced in any form whatsoever without expressed written permission from Tom Schmidt.

E-mail: tlsch@journalofthemollyb.com
Web Site: http://journalofthemollyb.com

Dedication

I would like to dedicate this volume of The Building of a Dream to all the people who have supported me and helped me along the way. It's been a very long haul and I have had a lot of support from friends and family along the way. Without them I don't think I would be able to complete this project. Your support means more to me than you will ever know. Thank You All!

Table of Contents

	Page
Swim Platform Bracing	1
Swim Platform Ladder	2
Rub Rail Cap	2
Saloon Paneling	3
Galley Cabinets	6
Fwd. Cabin Paneling	12
Head Paneling	13
Range Hood & Microwave/Convection Oven	14-15
Raised Panel Cabinet Doors	16
Stateroom Construction	20
Forward Cabin Flooring	24
Water Heater & Galley Sink Installation	26
Washer / Dryer Installation	28
Engine Control Cable Installation	29
Windshield Washer System	32
Salon Cabinets	34
Air Horn System	38
Flooring Installation	43
TV Cabinet	45
Emergency Bilge Pump System	46
Engine Exhaust Hose Installation	48
Fire Suppression System	52
Radar Mast Mount	54
Mast Step Hinge & Boom Swivel	57
Securing Labels to panel and valve handles	61
Shelves for the steering locker	63
Isolation transformer	65
Mast	66
Mast winch mount	69
Mast & winch wiring	71
Salon window valances	72
Boat deck hand rails	74
Boat deck ladder rails	76
Boat building tear down	78
Boat transport	81
Preparing boat for launch	87
Launch	88
Source List	91
Reference List	92

Introduction

This is Volume IV of the series "The Building of a Dream, Journal of the Molly B". It has been a long time coming but it's finally here. Volume II was published in 2007 and Volume III was published in 2011. I was hoping volume III would be the last volume before christening my boat. That doesn't seem to be very likely at this point in time. I have been working for the past four years trying to complete my project and I'm finally ready to put the Molly B into the water. All together it has been 11 ½ years since I started my boat building project. Not all of those years have been spent actually working on the boat. I have had a couple years where I was replacing my boat building after the hurricanes tore it down, and a few summers when I had to recover from major surgeries. So while the 11 ½ years seems like a really long time to devote to a project of any kind, I have actually been working on my boat for around 8 years give or take six months. When I look at it that way, it doesn't seem to be quite so long. Even 8 years seems to be a long time to devote to one project. It has been rewarding in many ways and the closer I got to completing my project the more personal satisfaction I got from all the years of work.

People have asked me every day how long it is going to be before I put her in the water. I used to say "Maybe another Year" but after saying that for the three or four years, I simply replied "When it's done". I have invested every dime I could come up with in this project, and now I'm at a point where the last of the items I needed to complete my project are purchased and installed. Most of which were major items that cost more than I could come up with all at once so I had to save up the money to purchase the things I needed and as a result it took a while longer to complete.

I have asked myself many times why I took on such a big project. Some call it a labor of love others just call me crazy. I decided to build this boat because I wanted a boat I can live aboard and enjoy where ever I want to go with it. When I started I had no idea that it would take me over 11 years to complete. I thought maybe three or four years and I would be cruising on my boat. I had no idea what I was getting into when I started this project. I've had to learn a lot as I have worked on my boat. I have had to do a lot of research to figure out how to do things when I had no idea at all as to how to do it. People ask me how I knew how to build this boat and I have to say I learned as I went along. That also takes a lot more time than doing something you already know how to do. Many, many times I have had to scrap what I was working on and start all over because it just didn't come out just right. I have probably built this boat twice already with all the parts I have thrown away. That too takes up a lot of time so all in all I think I have done pretty well with respect to how long it has taken me to get her ready to launch. I am very proud of my accomplishments to this point and can only say that it won't be long now before I put her into the water.

If you are contemplating starting a project like this, don't be discouraged when the years fly by and you still aren't done. Take your time, stick with it, and it will eventually all come together.

Welcome to Volume IV of the Journal of the Molly B!

This will be the last volume of my journal before the launch day. It's been a long hard road to this point but I can see the light at the end of the tunnel, finally!

Wednesday, March 02, 2011

Now that I have my swim platform completed I have started working on the braces. To the left is a picture of one of the braces in the vise formed and ready for polishing. This is stainless steel angle and it's really rough so I have to polish it to make it look good. That takes about two hours per brace so it's really time consuming but well worth it when it's done. Below is a picture of two of the braces shined up and two ready for polishing. Quite a difference. I worked all day today, and I have them all done except one so tomorrow I should finish them up. Then I have to drill them for mounting screws and then install them. Hopefully I can get all that done in the next day or so.

I have finished polishing my braces and I have started installing them. Below is a picture of the braces on the top of the swim platform. There are also three more on the bottom of the platform. I have them all installed and bedded in 3M 5200 sealant. The picture to the right below shows the three braces under the platform.

I've also been working on my ladder from the aft deck down onto the swim platform. To the left is a picture of the ladder. It's only two steps down, so I don't really know if you could call it a ladder but that's what it is I guess. Below is another view of the ladder.

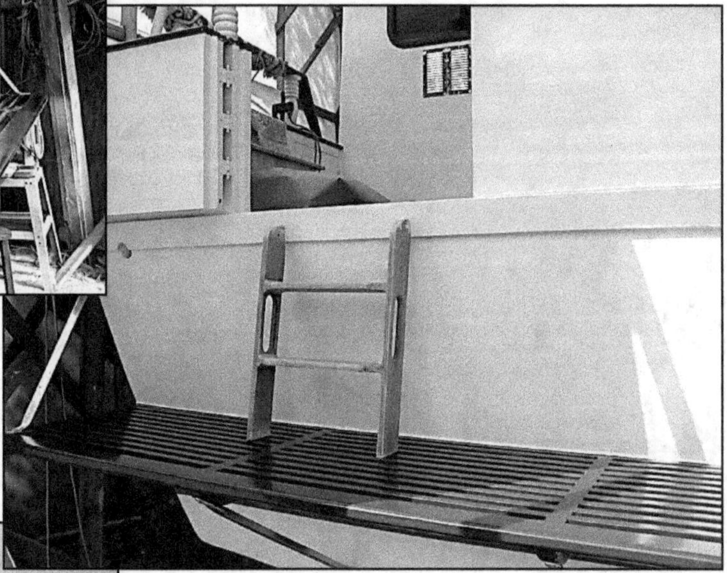

I have also completed the installation of my rub rail. Woody gave me a hand and we got it all installed in two days. Below is a picture of the rub rail along the starboard side. This stuff is made out of a hard rubber material and it was really hard to flatten it out to install it but we heated it up a bit and that helped. I had to make the end caps to taper the rub rail down and end it just aft of the anchor chain locker. I used a piece of pvc pipe and cut it down and tappered it as you can see in the picture to the left.

Friday, March 11, 2011

I finished my ladder down to the swim platform. To the right is a picture of the ladder finished and mounted to the transom. I made some mounting brackets out of stainless steel and bolted the top of the ladder to the transom. The bottom is screwed to the swim platform with two little angle brackets. I put some non-skid on the steps of the ladder and painted it with the topside paint I used on the hull. The swim platform was the last thing I had to build on the outside of the boat. So, the outside is all done now and I'm working on the inside.

I've spent the past few days insulating the overhead in the saloon and putting ceiling panels up. I'm using some vinyl covered ceiling panels for the headliner. They are textured and look nice. To the left is a picture of the panels I have installed so far. I'm nailing the panels around the perimeter along with gluing it up to the boat deck floors. I have the panel propped up with some supports as you can see in the picture. I leave the supports up overnight so the glue has a chance to completely set up. This makes for a slow operation when you can only put one panel up per day, but I'm finding other things to do between panels. When I get them all up I will strip the seams and put a grab rail up down the middle of the saloon to cover the center seam.

Saturday, March 19, 2011

We have finished the ceiling panels and the insulation of the side walls and we have all the paneling cut and fit. Below are a few pictures of the paneling. I have found some real nice looking crown molding to put around the top of the paneling. It is about three inches wide and has a rope pattern inlayed into the molding. It looks very nautical and I think it will really set off the paneling from the ceiling panels. I plan to stain and varnish the bottom three feet of the paneling and then put a chair rail around the room and then paint the top part of the panels white so it won't be so dark in there. You can see the line in the picture to the left that indicates where the chair rail will go. To the right is another view of the paneling. To the left is a picture of my pantry which I spent most of today working on getting it paneled inside and out. I almost have it done and then I can start taking the panels down and staining and varnishing them.

Tuesday, April 05, 2011

To the left is a picture of some of my panels on the work bench. I am staining them and getting them finished before I put them back up again. As you can see on the first panel, it is only stained at the bottom and will be painted on the top portion. Below is a picture of some of the panels varnished with the top portion painted. I found an off white to use as a base coat and then I sponged a little darker color over that which gives the panel a textured look. Below is a picture of the panels put up in the saloon. You can also see my grab rail I made installed down the center of the saloon. This black and white photo doesn't really do it justice. The headliner is white while the top portion of the panel is off white with the textured look, it really looks nice. I have all my trim stained and varnished as well. To the right is a picture of my trim all stained and varnished. I ran out of varnish and I'm waiting for my order to come in before I can complete the panels for the port side. In the meantime I am installing some of the crown molding along the starboard side. I have the casing done for the back door so as soon as my varnish comes in I can install that and then install the chair rail along the port side.

Saturday, April 09, 2011

I've been working on some of my molding. In the picture to the left you can see some of my crown molding is installed. I can't go too far because I still don't have the port side paneling put up. I did get my varnish yesterday so I got some of my panels varnished today. I also put some chair rail up so I could see how it was going to look. In the picture below you can see my chair rail installed along the starboard side. You can also see one of the headliner battens installed and a couple of bulkhead battens installed. The only thing left in this aft corner of the saloon is the window trim. That will really give it a finished look. These pictures make the stained paneling look really dark, almost black, but they are actually a dark mahogany.

Tuesday, April 13, 2011

My varnish has arrived as I mentioned and I have finished the portside panels. Woody and I put them up today and I got started on the crown molding around the aft of the saloon. I also got the chair rail installed along the aft bulkhead as well. Now I need to lay out the galley cabinets so I know how far to run the moldings along the port side. To the left is a picture of the port side showing the sliding doors along the bottom of the paneling. The sliding doors allow access to the storage areas under the side decks. It will be difficult to access these areas because of furniture being in front of the panels but it is storage none the less so I decided to put the doors in to access it even though I will have to move the furniture to get to it. I have also completed the crown molding along the starboard side all the way to the front of the saloon. The top left picture on the next page shows this molding all installed. It's really challenging trying to cut the crown molding to fit the bulkhead when they aren't at right angles to each other.

Below is a picture of the panels along the port side. I have them all installed and I'm starting to cut and fit the

molding. Below is a picture of the paneling in the Galley area. I have removed the varnished section at the bottom of the first panel aft of the galley window because the cabinets will cover this area and the varnished section extends above the counter top of the cabinets. I will paint and sponge this section as I have done with the rest of the panels in the galley area.

Tuesday, July 12, 2011

I'm back to work after my knee surgery finally. Actually I have been working for the past couple of weeks but am just now getting around to getting my journal up to date. I have been working on the trim for around the windows in the saloon and laying out my cabinets for the galley. I have all of the window trim made and the corner pieces are sanded and stained. I just need to sand and stain all the straight pieces and then varnish them. In the meantime, I have been working on my galley cabinets' layout. This has been a real challenge since I have very limited space for all the appliances I need to get in there. The picture to the right shows the base for one of the cabinets on the workbench. I can't kneel down yet so I needed to get it up where I could work on it so I moved it to the workbench to glue and screw the whole thing together. I have one

of the end pieces installed just to see if it would fit correctly.

To the left is a view of the base I had on the workbench installed in the galley. I have added some of the upright parts of the cabinets in this picture. It doesn't look like much in this picture bur it is a work in progress and will take shape as I get more of it put together.

Thursday, July 14, 2011

I've been making some pretty good progress with my cabinets. Below is another view of the basic layout for the lower cabinets. I spent most of today gluing and screwing some of the framework together. I need to get a few more pictures of what I have done to date.

Saturday, July 16, 2011

My cabinets are coming along pretty well. Below is a picture of the outside panels installed. There will be a cabinet door on the cupboard on the left end of the cabinets. Below is a view of the cabinets looking aft. You can see the opening in the center of the cabinet for the trash compacter in the picture to the right. To the right of the trash compacter there will be a row of drawers. The opening on the left of the cabinet is where the clothes washer/dryer will be installed. The counter top stove will go right above the drawers.

Sunday, September 11, 2011

I haven't made any entries in my journal for some time because I just got out of the hospital ICU. I came down with a bad case of pneumonia and had to spend two weeks flat on my back. I'm home now but I'm very weak so it will probably be awhile before I'm able to get back to work.

Saturday, October 15, 2011

I'm back to work finally! I'm still working on my galley cabinets. I have been building drawers for the past week or so, and I have them pretty well constructed. To the left is a picture of the cabinet with the drawer slides installed. I have also completed the installation of the water faucets for the

clothes washer and the electrical for it. Below is a picture of the hookups.

I built the drawers on the front workbench. Below left and right are a couple views of the drawers. It sure is good to be back to work on the Molly B again. It's been a long time and I'm still not quite back to normal but I'm making some good progress.

You can just make out the silverware drawer I made in the picture to the left.

Sunday, November 20, 2011

I'm still working on my galley cabinets and they are coming right along. Here are a couple of views of the cabinets after I have them all sanded, stained and varnished. I also sealed the inside of the cabinets with primer just so nothing will wrap if it gets wet or anything. Below is a view of the inside of the cabinets. At this point I'm ready for the counter tops and I'm having them made instead of me trying to laminate them. I did that once when I built my house and it turned out good but I want a fancy beveled edge on these so I thought I better have a professional do that. Below is a view of my counter tops looking forward. Notice the little spindle trim I'm going to install on the top of the counter behind the stove top in the picture to the left. Below is a view looking into the galley from the end of the cabinet. You can see the pattern in the counter top a bit better in this view. We picked a granite type pattern and it really looks great with the mahogany stained cabinets. I have the stovetop and the sink set into the countertop just to see how they fit. I may have a problem with the sink since I am going to use sliding doors below the sink and I think the mounting brackets for the sink may interfere with the sliding doors.

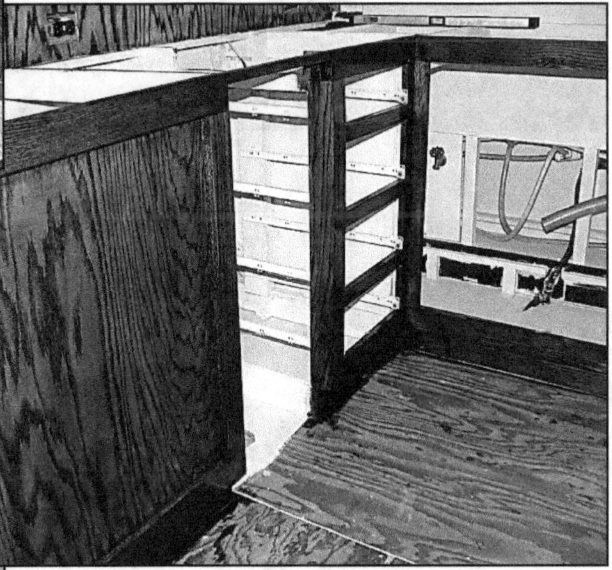

To the left is a picture of the cabinet tops from another angle. You can see the drawers installed in this view as well as the stovetop installation. I have the upper cupboards all framed up and glued and screwed together. All I have to do up there yet is panel them and install the counter top lighting which goes under the upper cupboards. Then I can get them all stained and varnished as well. I just really have to be careful not to mess up the counter tops while I'm working on them. I probably should have waited until I had them all done to install the counter tops but I really wanted to see how they were going to look so I went ahead and installed them.

Wednesday, December 14, 2011

I'm still working hard on my galley cabinets. I have the exhaust duct for the range hood completed. To the left is a picture of the exhaust duct where it goes through the partition between the convection oven compartment and the outside cupboard. Below is a view of where the duct comes out in the outside cupboard. If you look closely you can see the opening in the outboard bulkhead where the exhaust goes to the outside in the upper right side of the picture. To the left is a picture of the duct to the opening. There is a cover that goes over this duct to insulate the exhaust from the cupboard contents.

I have completed the cupboards; they are all stained and varnished so I have started working on the cupboard doors. I'm trying my hand at building raised panel doors for the cupboards. I've never done them before, so this aught to be interesting. I purchased two cutters for my shaper, one for the stiles and rails, and one for the panels. The cutter for the stiles and rails is a reversible type cutter so it cuts both the stiles and rail ends depending on the way you stack the cutters on the spindle. A bit confusing but I figured it out and got all the stiles and rails cut. To the left is a picture of all the cabinet doors all cut out and ready for assembly. Below is a picture of one of the doors assembled to show what the raised panel looks like. They really turned out great although it is a lot of extra work doing the doors this way.

Tuesday, December 27, 2011

To the left is a picture of a few of my cabinet doors installed. I have finally finished them and I'm in the process of getting them hung. I just finished hanging the rest of the doors today, but I didn't have my camera with me so I will have to get some pictures of the others tomorrow. I have also been working on getting my windows trimmed out. I got all the trim stained and varnished and I have all the windows done now. To the right is a picture of some of the windows with the

trim installed.

Wednesday, December 28, 2011

To the left is a picture of the pantry doors. These were fun to build and actually turned out pretty well.

Below is a picture of the cupboard doors on the cupboard next to the refrigerator. These two doors are hinged together in the center so there is only one pull and the doors open like a by fold door. This way the outer door isn't in your

way to get at what's in the cupboard. I put two little brackets on the edge of the inside door to hold it against the cupboard when it's closed. The outside door has a catch which actually holds both doors closed.

Wednesday, January 25, 2012

I have moved down into the forward cabin area. Here are a couple of pictures of the paneling I have done so far. To the left is a view looking down the stairs into the forward cabin and to the right is a view of the storage compartments all paneled.

I'm using every square inch for storage, but that means I will have to make lots more cupboard doors.

Friday, April 06, 2012

Boy, talk about being behind on my journal! I'm still working down in the forward cabin area, and I've made some pretty good progress so far. To the left is a picture of the paneling on the engine room bulkhead. If you look closely you can see the engine room door in place. I don't have it installed yet but I wanted to make sure it would fit the opening so I stuck it in there to check the fit. I'm going to have to put a threshold under the door so it will seal along the bottom edge yet and then I'll be ready to install it. Below is a picture of some of the

paneling I have completed in the head. As you can see, I have the wiring done, and the sink installed. The plumbing was a real challenge but I managed to get it all hooked up. The big opening in the center of the picture is for the medicine cabinet. In the picture to the left you can see the openings for the cabinets above the sink. I pretty well have all the cabinet work done in the head now and I just need to get it painted. As much as I love to paint, it might be awhile. As soon as I get some of the painting done, I will have to get some more pictures and get the rest of this operation caught up.

Sunday, April 08, 2012

I have managed to get quite a bit of the painting done down in the forward cabin area, at least in the head. I have finished all the primer in

the head and now I just have to sand it all down and put the final coat of paint on in there. I have a couple of pictures of the head with the primer on it. To the left is a picture of

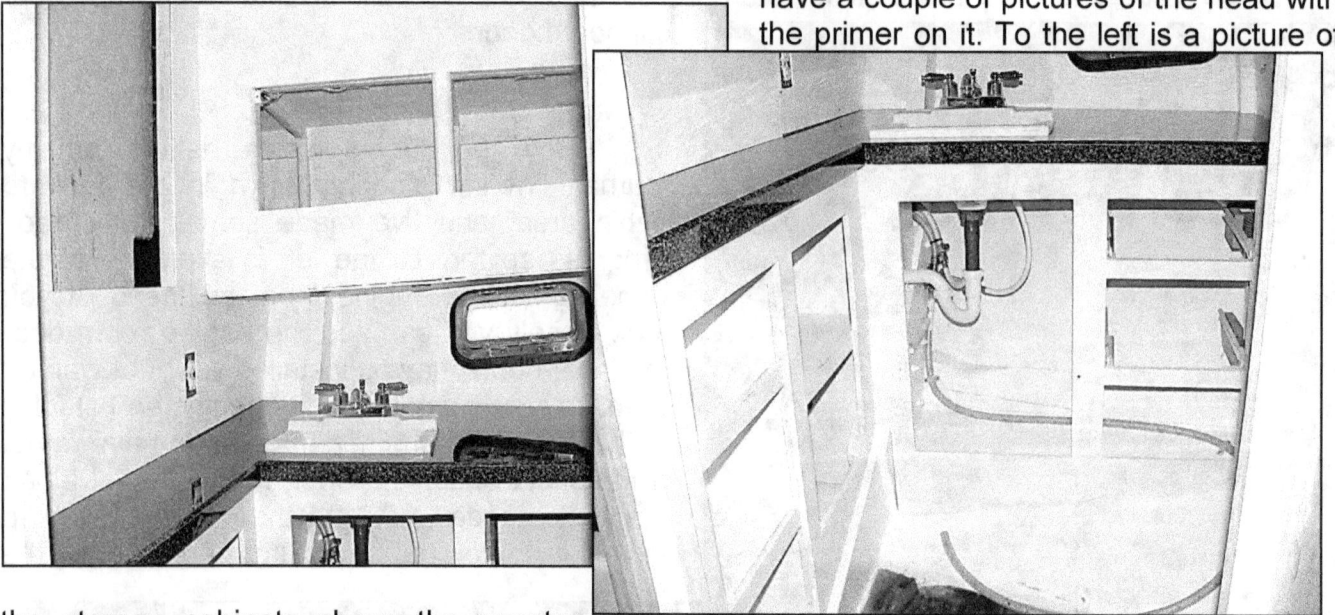

the storage cabinets above the counter top in the head. Below is a view of the lower portion of the cabinets in the head. To the left in this picture are several shelves which will have regular cupboard doors on them. To the right in this picture below the sink is a storage compartment which will have shelves and to the right of that there are a couple drawers. We should have plenty of storage in the head.

I have also received my convection/microwave oven and range hood and have them installed in the galley. To the left is a picture of the range hood and the oven installed. I have the trim kit on order for the oven but it hasn't come yet. I will be able to finish the installation as soon as it arrives.

Sunday, April 15, 2012

I received the trim kit for my microwave/convection oven the other day, and I have it all installed. Cross off one more item on my list of unfinished jobs. They sure seem to pile up, as there is always something else that needs to be done or ordered before you can finish what you started. I also finished fitting my headliners in the forward compartment. This was a real challenge as you can see in the picture to the right; there is nothing but angles to cut with these headliner panels. They need to be removable because all my wiring runs under them and I may need to get into that control closet again.

Friday, April 27, 2012

Here is a picture of my Trim Kit for the convection/microwave oven. That

really finishes it off. Below is a picture of my head. I have it all painted and the drawers installed. I have also installed the medicine chest, picture to

the left below. All I need now is the cabinet doors and the head will be done.

Now I'm back on the mend again. I had my right foot operated on last Tuesday so I will be laid up for 10 to 12 weeks. This is the foot I broke when I fell off the scaffolding back in '04. It was never set properly and it was really starting to give me a lot of trouble so I had to get it fixed. What ever you do, don't let anyone operate on your foot! Very Painful!

Saturday, June 02, 2012

Well, I'm getting nothing done on the Molly B. I've been sitting here in my recliner for the past five weeks waiting for this foot to heal. I have another five weeks to go before I can get this cast removed, during which time I have to stay off my foot completely. Once the cast is removed, I hope I will be able to get back to work; I'm going crazy just sitting around, and I really have a lot to do.

Sunday, July 01, 2012

I'm nearing the end of my convalescence, just one more week to go and I hope I can get this cast off. I don't know what's in store for me after that but I hope I will be able to walk so I can get back to work. I ordered my anchor windless so I can get that installed when I get back to work. I need to get the wiring installed for that before I can finish the stateroom because the wiring has to run through there to the windless. Then I guess I will have to start making some more cabinet doors. I need another fifteen doors for all the compartments in the forward cabin area and I still need four more doors for the galley cabinets. Two under the sink and two over the fridge, then all the cabinets in the galley will be done.

Wednesday, August 01, 2012

Here it is August 1st already. I did get my cast off almost three weeks ago, but the Dr. said I need to stay off my foot for another two or three weeks. That time has expired, and I still can't walk on my foot. I can however put some weight on it now without a too much pain so maybe in another few days I can get back to work. I ordered my water heater the other day, so I will have that to install in addition to everything else I have to do. The work is really piling up and I'm really getting anxious to get started on it. It's really hard sitting around when I'm getting so close to the launch date!

Friday, August 17, 2012

I'm making some progress with this foot of mine all be it very slowly. I have started to walk on it a little bit now but it's still a little painful. I will be starting physical therapy shortly to try and get some range of motion back in my ankle because it's really stiff right now, and then I hope I will be able to get back to work.

Sunday, September 09, 2012

I'm finally back to work although only a couple of hours a day. My foot is finally getting well enough to be on it for a bit longer. I have been working on my cabinet doors for the past couple of weeks and I have the last 24 doors all cut to size with

the stiles and rails all profiled. I started working on the raised panels today and have five of them cut to size but still need to profile them. To the left above and to the right are a couple pictures of the stiles and rails that I have all cut out.

Saturday, October 06, 2012

After another short break, I'm back to work again on my cabinet doors. My Mom turned 92 last month so I had to take a trip up north to wish her a happy birthday and attend my 50th class reunion. It sure was good to see some of my classmates again.

Now I'm back to work on my cabinet doors. I have been working on my raised panels for the doors for the past few days. I have them pretty well all cut to size and I have most of the panels glued together that are wider than a 1 x 12. There are about eight or so panels that I had to glue together a couple pieces of oak to make them wide enough for the door. I still need to get the profile cut on all the panels and then I can start assembling them. I sure will be glad to get finished with these doors. It's not hard work but it takes a long time to get them all cut and fit and then put together.

Monday, October 30, 2012

To the left is a picture of some of my doors all put together. This is only about half of them so I still have a ways to go before I'll be ready to start sanding and staining. I finally completed the assembly of all my doors and have them all sanded. I started staining them and you can see the stained doors in the picture above. Black and white pictures just don't show how beautiful they turned out so I guess I'm going to have to put out a version of my journal in color. That might be a bit expensive but I think I will have to look into it. The bottom right picture on the previous page shows my doors with the first coat of varnish on them. I'm going to put

three coats on the front sides and maybe just one or two on the back sides. This should seal them up as well as providing the outside surfaces with good protection and durability. To the left is another picture of the rest of my doors with a coat of varnish on them. Sure are a lot of doors.

I have all my doors varnished and I have started hanging them in the forward cabin and head. Below is a picture of my doors installed in the companion way going down into the forward cabin. As you can see in this picture there is one door missing. I messed up when I made this one and put the angle on the bottom instead of the top so I have to make that one over again. Not bad though, only one out of 24, I hope!

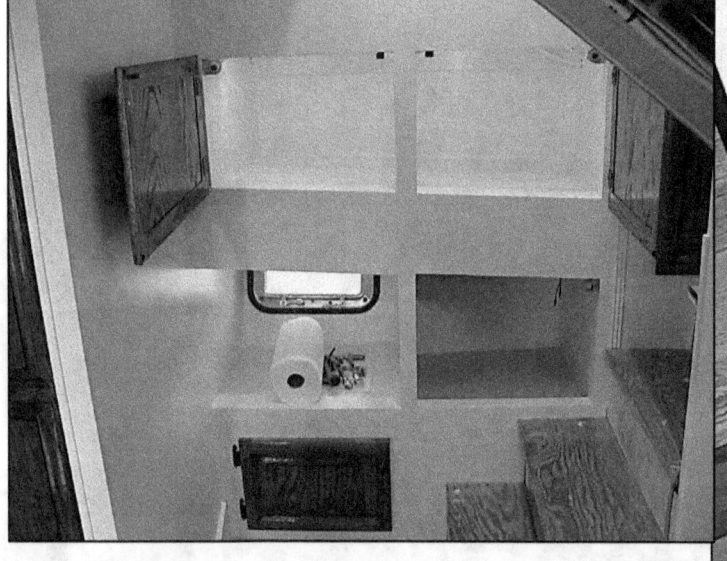

Above is another view of these same doors but with the top doors open. You can see that the storage compartments are lit up when the doors are opened. I have a switch on each of the door jams that turns the light on and off when the door is opened or closed respectively. I have installed this lighting in almost all of my storage compartments and lockers in the forward cabin area as well as in the stateroom. Many of these little compartments are so small that they are really dark inside when you open the door so I thought it would be better to be able to see what's inside and I needed a light to do that so I put them in. It was a lot of extra work but really worth the effort.

To the left is a picture of the two big doors on my hanging locker. You can see the little compartment below the doors in front of the locker. This is going to be a dirty clothes hamper and a bench type seat to sit on while dressing. I have to make a lid for this yet but that is what it is for.

Above is a picture of the inside of the hanging locker. As you can see, it is also lit with a couple of interior lights. I put a light in the locker itself as well as in the two storage compartments behind the hanging locker. You can see a hatch on the bottom of the locker, this is another storage compartment. This one has a lock on it to store more sensitive items.

To the left is a picture of my head. I have all the doors installed in here and except for the flooring and setting the head itself, it is pretty well done in here. As you can see, I put two mirrored doors above the sink. There is also a large mirror on the front of the medicine cabinet. I also have my engine room door installed and I'm waiting for the latch to come in and I will be done with that.

The forward cabin is almost done, and next I move into the stateroom where there is a lot of work to do to get it finished.

Monday, November 12, 2012

I've moved into the stateroom and I'm making some good progress in there. I have removed my port lights and have the hull sealed. I'm waiting until I finish the painting and then I will reinstall them. I just have one little area that needs a third coat to cover that area a little better. I have finished painting all the headliners and I have added six more down lights in the headliner above the bunks for a little extra light in there. I have the main overhead fixture installed already. I also have my paneling fit and varnished around the trunk cabin walls so as soon as the varnish sets up a bit better I can install them. Today I made my template for the plywood that goes over the framework of the bunks. I laid the pattern out on the plywood and it takes up the whole sheet and there isn't much scrap left over. I have also finished installing all the drawer slides for the drawers under the bunks. I have two of the drawer fronts made with six more to go. There just isn't much to take a picture of yet bet as soon as I get things installed I will get some pictures taken.

Sunday, November 25, 2012

I finally got some pictures of what I'm doing in the stateroom. To the left you can see the paneling around the trunk cabin installed. You can also see the little LED down lights I have installed along the overhead above the bunks. I've finished the painting of the frames and

topsides and I have my port lights installed. I also have the portside plywood cut and fit for the top of the bunk as you can see in the picture above. To the right is another view of the bunks. You can see I have the doors installed for the forward hanging locker and the AC outlet wired and installed. This is an outlet I put up there just in case I might want to put a small TV in the stateroom someday. I also have the coax run up there for the antenna signal.

To the left is a picture of the aft bulkhead of the stateroom. I just wanted to show some of the wiring I have completed to date. You can see some wires hanging down along the left side of the picture which are the wires for the reading light that will be mounted on the paneling above the bunk on this bulkhead. You can also see some of the AC wiring for the outlets alongside the bunk.

To the right is a picture of the bulkhead on the starboard side. The wiring here is much the same except there is wiring for two switches and the wiring for the anchor winch. You can see the AC wiring for the outlets cross the DC wiring for the anchor winch at 90 degrees.

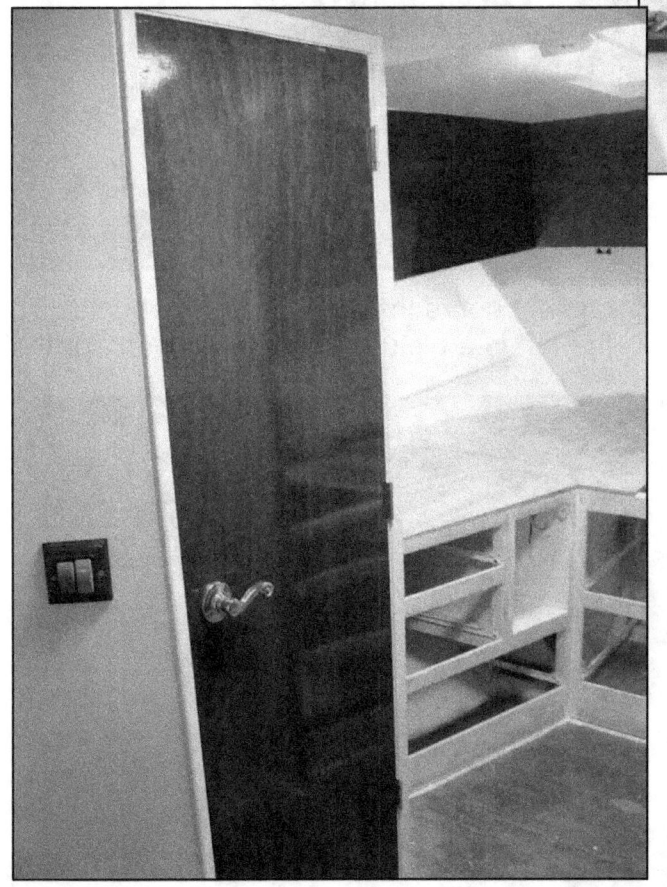

To the left is a picture of my door for the head. I finally got it all put together and hung the other day. It actually fit pretty well. I only had to sand down a couple of spots a little to get it to close properly. So, I guess it turned out pretty well.

Today I spent the day working on the other plywood panel for the starboard side bunk. I got it all cut and fit by using the port side plywood as a template. Then I cut out some access hatches forward of the drawers where there is a large open space under the bunks. This will make a good place for storage of things you don't need to get to very often. I got the hatch openings all cut out and the battens installed to support the hatch covers. Both panels are ready for primer so I will get that done tomorrow.

Monday, November 26, 2012

I got one side of my bunk panels painted this morning. To the left is a picture of the two forward sections. These are the panels that have the hatch openings in them for access to the storage compartment below the bunks.

Tuesday, December 04, 2012

I'm still working on the stateroom. I am making some good progress up there but it is taking some time. It's hard to work in this somewhat confined space. Now that I have the panels installed on top of the bunk framework, I have to reach over the bunks to work on the shelves alongside of each of the bunks. To the right is a picture of the stateroom showing the drawers and cabinet doors installed under the bunks. To the left is a picture of the framework for the shelf along the side of the starboard bunk. This will provide a straight surface for the mattress to butt against instead of having a gap between the mattress and the side of the hull. To the right is a view of the port side framework. As you can see, I don't have it completed yet. I still have a couple pieces to cut and fit forward and it will be done as well. Then I have to cut and fit the front panels of the shelf and cut and fit the top of each shelf. You can also make out the access panels in the plywood panels on top of the bunks that provides access to the storage compartment under the bunks, just forward of the area of the drawers. There is an access panel on each side of the bunks.

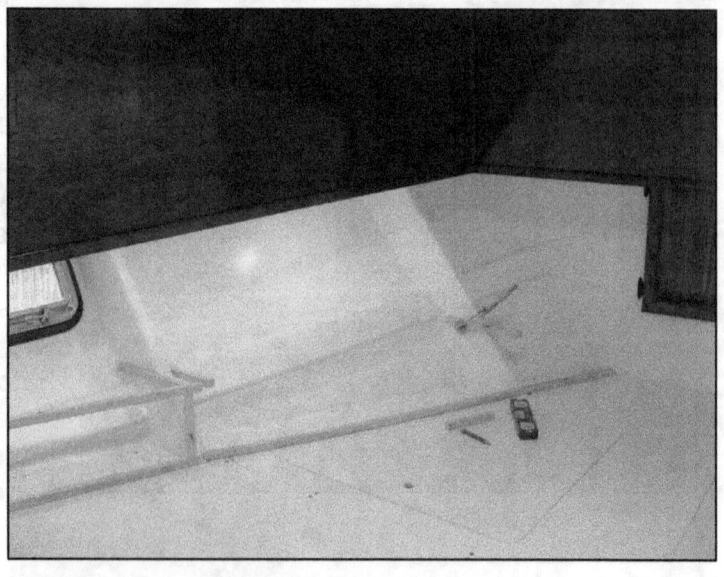

Thursday, December 13, 2012

I've finally completed my shelves for the stateroom. To the left and below are a couple pictures of my shelves on the workbench where I have them all varnished and ready for installation. You can also see the galley rails in the picture to the left which are the rails with the little spindles in them that will go along the edge of each shelf. These will provide something to prevent objects from falling off the shelves. These are the same as the ones I used in my galley. In the picture to the right you can

see the two panels that go over the two forward frames in the stateroom.

Saturday, December 15, 2012

I have my shelves and paneling installed. In the picture to the left you can see the shelf all put together and the panel installed just forward of the port light. You can also see the galley rails which I still need to install. I wanted to wait with these until I finished the caulking around the panel. I have the caulking finished on the starboard side but ran out before I completed the port side. I only have a foot or so to go on the port side and I will be finished. To the right is a picture of the shelf under the port light. You can see that the paneling stops just forward of the port light in this picture. I didn't want to try to frame in around the port light so I just stopped the paneling at the frame just forward of it.

Tuesday, December 18, 2012

I've finally completed the shelves on each side of the V berth. As you can see in the pictures to the left and below, the galley rails are installed. This was the last thing I had to do with the shelves. I have also completed the caulking in the stateroom so except for the molding around the front of the bunks and the floor, the stateroom is done.

As you can see in the pictures to the left and below, I have started on the flooring in the forward cabin area. The picture to the left shows the floor planking in front of the shower. Along with the strip in front of the door to the head, this took me all day to get installed. There just isn't anywhere on this floor that is straight. It's all angled and there are a lot of objects to notch around which really takes a lot of time to get right.

Thursday, December 27, 2012

I have finally almost completed the forward cabin area. I have the flooring completed, the quarter round trim all installed, the toilet is all hooked up and installed and I have one of the two hatches fit and trimmed out. The only things I have left is to fit and trim out the second hatch, and touch up some paint in a few places, and install the molding in the front of the bunks. It has been a long time but I am finally almost done!

Friday, December 28, 2012
Here are a couple pictures of my head installation. This was quite a project which took me a little over three days to complete. There are a lot of connections to make behind the toilet most all of which had to be modified to connect to the plumbing and wiring I had installed for it. I spent almost half a day just running around looking for parts to make the modifications with, but I finally have it completed. As you can see in these pictures, I have my quarter round all installed and today I finished touching up my paint. You can also see the hatch I have trimmed out in the picture above. I just have the aft hatch to complete and install the molding on the bunks, and I will be completely done with the forward cabin.

Next I need to put some water in the boat so I can test my head installation to make sure there are no leaks and the wiring is working properly. I can't do that until I finish installing the water heater and the faucet on the galley sink. Then I can check all my plumbing connections for leaks. I sure hope the water heater installation goes better than the head did.

Monday, December 31, 2012
Happy New Year to all!

Another year has passed and I'm still working hard trying to finish my boat project. I have the forward cabin done except for the trim around the aft hatch cover. I have moved up into the galley and I'm working on installing my water heater. It's going pretty well so far, I have all the water lines ready to hook up and the wiring is also ready. I just need to build my LP gas system and then I can hook it all up.

Tuesday, January 01, 2013
I've finished the trim on the aft hatch as you can see in the picture to the left.

I've also completed the stateroom. The only thing I had left to do in there was install the trim around the edge of the bunks. The picture to the left shows the trim installed.

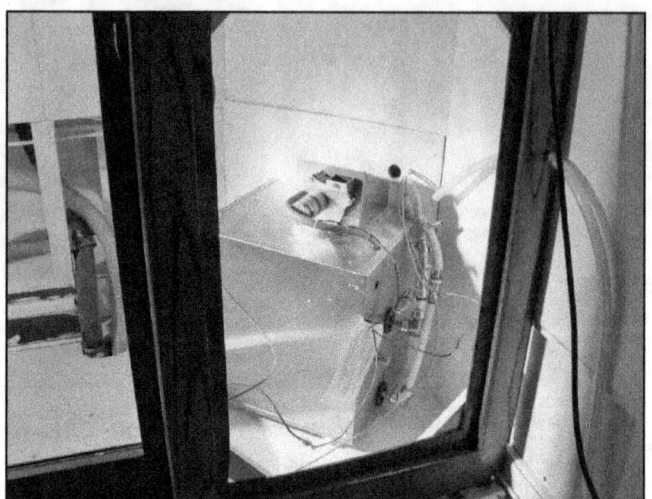

In the picture above you can see my water heater installation. This is a little on demand water heater and it just fits under the galley counter. The picture below gives you a little better idea of just where the water heater is located. It is just to the right of the sink and just aft of the shower. It should provide plenty of hot water to all the taps with very little wasted water due to its close proximity to all the taps. I still need to run the LP gas line from the propane locker to the water heater and the cook top and then I can complete the hook-ups to the water heater. I have the water lines cut to length but I need to get behind the panel that's behind the water heater to run the gas line so I need to get

that done before I can finish the hook-up.

Wednesday, January 09, 2013

Here is a situation that I don't know how to resolve. I can't get the sink to fit into the cutout in the countertop because the cutout is too big. As you can see in the picture to the left, the little hold down bracket is not hitting the underside of the countertop with the sink centered in the opening. The picture on the next page shows the bracket on the back of the sink

and it is the same as the front one. It won't reach the underside of the countertop. I have tried to bend the little brackets but all that did was mess up the threads in the hole for the bolt. I had to rethread the hole to salvage the bracket. If this isn't bad enough, take a look at the back of the small sink in the picture below.

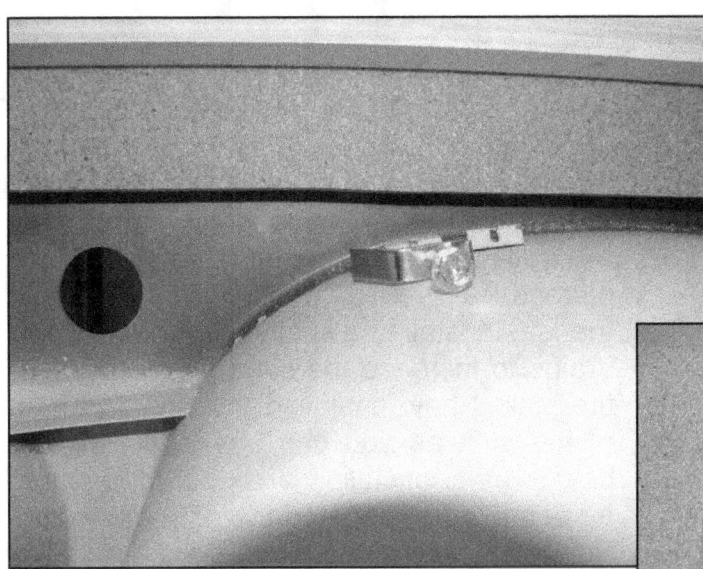

This bracket is several inches from the underside of the countertop. I have contacted the company that installed the countertop and they said most people don't wait a year to install the sink and implied that I must have purchased a different sink and that's why it won't fit. So, I now have to try to figure out how to make the cutout hole fit the sink instead of the other way around. If you're in the state of Florida, and buy a counter top from Home Depot, make sure they us an installer other than Ruhl Enterprises of St Petersburg, FL. They do shoddy work and won't stand behind their workmanship.

Saturday, January 12, 2013

Here are a couple pictures of how I fixed the oversized hole for the sink. I just added a couple blocks under the countertop to extend it so the clips would reach them to secure the sink. Now I can finish the sink installation. I spent the day today running around getting fittings for my faucet so I could install it. I finally found what I needed and installed the faucet. Now

I just need to find someone that has a 1-1/16th inch drill bit so I can drill a hole in the sink to install the little liquid soap pump on the sink.

Saturday, January 26, 2013

I'm making some good progress in the galley. I have my sink all installed and plumbed as you can see in the picture to the left. I installed a shelf just below the sink drains to increase the storage space under the sink. I have installed the doors in front of the sink as you can see in the picture below. I also installed a couple of louvered air vents in the panel to the right of the sink

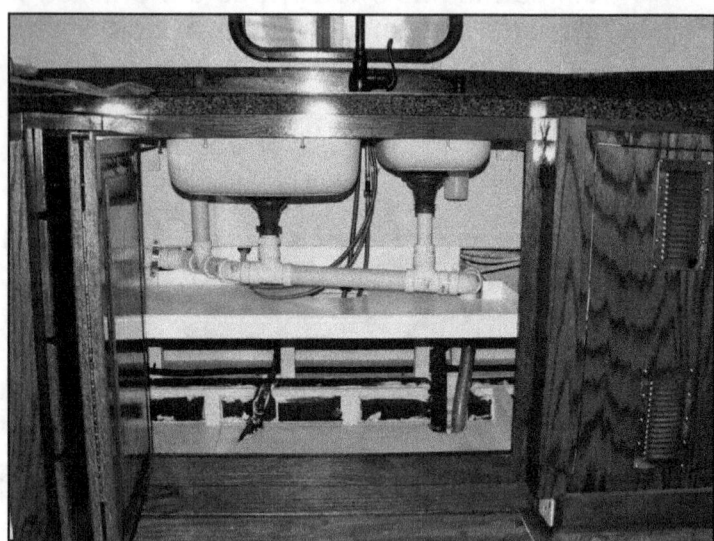

for the water heater behind it.

I have completed the installation of my wash/dryer. When I built the cabinets, I used the dimensions given for the washer and when I installed it I only had about 1/8th of an inch to spare. A pretty tight fit, but it did fit. To the left is a view of the front of the machine. I put a drip pan under the machine to catch any small leaks that may occur. I also made a couple of hold down brackets that I put under the front leveling feet of the machine and screwed them down to the deck. In the rear of the machine I used a couple of angle brackets to keep the machine from sliding side to side. It is pretty well secured so it can't move around at all.

To the right is a view of the rear of the machine. As you can see in the picture, I had to run all the connections to the washer to the back of the cabinet in order to allow room for my trash compactor which will be installed in the just behind the washer. I had to extend the power cord and the discharge hose but everything else fit pretty well.

Saturday, February 09, 2013

I ended up going to a muffler shop to have an exhaust pipe made for my water heater. The one I purchased wouldn't fit over the 2 inch thru hull I am using for the exhaust so I had to have one made and expanded on the end to slip over the fitting. I got it installed today. I had to put some heat on it and bend it just slightly to get it to fit onto the water heater and the thru hull. I'll have to get some pictures taken so I can show you the exhaust pipe, it's really quite impressive. Here it is, below is a picture of my exhaust pipe. I have it all installed finally.

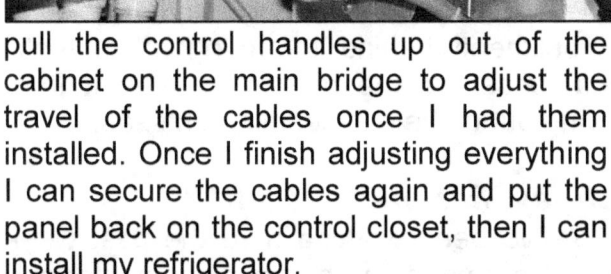

Tuesday, February 12, 2013

I've also been working on getting my control cables hooked up to my engines. I need to get them installed and adjusted before I can put my refrigerator in because the cables to the fly bridge go through the control closet that runs through the wall along side of the refrigerator. I needed to loosen all the cables to get enough slack to pull the control handles up out of the cabinet on the main bridge to adjust the travel of the cables once I had them installed. Once I finish adjusting everything I can secure the cables again and put the panel back on the control closet, then I can install my refrigerator.

To the right is a picture of the throttle control cable all installed and adjusted for travel. The mount adaptor is shown attached to the plate I had to make to bolt it to the engine. The cable is attached to the top of the mount and the push pull rod end

bearing is attached to the throttle lever on the injector pump.

To the left is a picture of the adaptor mount and control cable attached to the transmission. This is the starboard engine so with the control lever in the forward position the transmission is actually in reverse as it is in this picture. This provides the counter rotating propellers needed to offset any torque that would cause the boat to list one way or the other. Both transmissions are rated the same in forward and reverse.

To the right is a picture of my trash compactor installed in the cabinet in the galley. What you see in this picture is the inside of the compactor because I have the front panel removed. I have to make a

cabinet door panel to mount to the front panel of the compactor, and then it will look like the rest of the cabinets.

To the left is a picture of my refrigerator installed in the galley cabinets. I have the top of the refrigerator secured with an L bracket to the bulkhead behind the refrigerator. I still need to secure the bottom to keep it from moving out of the cabinet in rough seas.

Sunday, February 24, 2013

I've been working hard the past week on my door panel for the trash compactor and doing a lot of wiring on the bridge. I almost have everything done on the bridge except for my windshield washers and a few more wires to run and I will move back down into the saloon.

To the left is a picture of my door panel that goes on the front of my trash compactor. Below is a view of the back side of the panel with the metal panel laid out on the back side of the wood panel. The metal panel gets screwed to the back of the wood panel.

Below is a picture of my overhead console. I had a lot of wiring to complete up there. As you see, I have my VHF radio mounted and wired, my depth sounder is mounted and wired as well. I still need to go down into the engine room to install the transducer but the display is working. I also have all my switches installed and hooked up as well as my search light control panel. I also installed my horn and the switch to operate it. It's really hard to see, but the horn button is on the right side of the console. I also have the wiring for the horn switch run to the fly bridge so I can operate it from either bridge.

To the right is a picture of my remote monitor installed in my center console. This is a touch screen monitor which will allow me to operate my navigation software from the bridge that will be running on my computer at my navigation station. I don't have it hooked up yet because I'm waiting for an external VGA card to plug into my computer. I don't have a video output on my computer so I needed to use an external one so I could adapt the connection to a USB to plug into the back to my computer.

Sunday, March 03, 2013

I'm still working on the bridge. I have completed my windshield washer system and it really works great. To the left is a picture of my washer fluid holding tank. This is a 10 gallon tank, so I don't think I will be running out of fluid any time soon. This was about the smallest tank I could find so that's what I put in there. In the lower right corner of the picture you can see my little pump. I had my doubts that it would pump the fluid all the way up to the windshield but it does, and with plenty of pressure. You can also see the filler hose coming down in the left top corner of the picture. I installed the filler opening inside of the chart table so it is not visible unless you open the chart table top.

To the right is a picture of my shut off valve I installed in the hose going up to the wipers. This will allow me to turn off the fluid, when I have to remove the top of the bridge for transport, so it won't run back down from above and make a mess. I probably didn't really need this valve, but I thought it would prevent a major spill of fluid.

I have also completed the wiring on the bridge now. I had two more switches to install in the overhead console, one for the depth sounder, and one for the windshield washer pump. I just need to run the VHF antenna coax and the hailer horn wiring through the roof and I can close up the overhead console. I have completed installing the cables that run from my remote monitor on the bridge down to my computer which will be located at my navigation station in the saloon. I also have my hydraulic lines all hooked up to my helm pump so I'm almost ready to test that system for leaks and proper operation.

Friday, March 08, 2013

I've made some pretty good progress on my bridge. I have it almost all put back together. I have my ships wheel installed as well as the panel behind it as you can see in the picture to the left. Below is another view of my ships wheel. It's a big wheel but it looks good and it's real easy to turn the helm pump. I need to get some hydraulic fluid in the system so I can check it for proper operation and any possible leaks. I will need to get my hydraulic cylinder hooked up in the steering locker before I can do that though. I still have a couple of wires to hook up in my overhead console and an air hose to run up the

starboard corner of the windshield for my air horn and then I can close the rest of the bridge up. Then all I will have left to do up there is install the deck and get the cushions made for the bench seat.

While I'm waiting for parts to finish the bridge, I've moved back down into the saloon. To the left is a picture of the basic layout for the last of the cabinets I have to build there. In this view, you can see the dinette in the back corner of the saloon. This is the framework for the entertainment center and the bench seat for the dinette. The base frame in the foreground is for my navigation station desk. I have all the base boards laid out and I'm starting to work on the stiles or the vertical cabinet face pieces. Once again I'm working around a bulkhead that is curved and angled inward so building these cabinets is lots of fun!

To the left is another view of my cabinet framework. Looking forward you can see the frame for the navigation desk a little better. It's coming along pretty well, but I have a long way to go with these cabinets.

Tuesday, March 12, 2013

I've been working on my cabinets for the past few days, and I'm making some pretty good progress. Below is a view of my cabinets looking aft. You can see the bench seat in the back of the picture. I have the plywood cut and fit for the seat and I installed some little blocks on the underside to

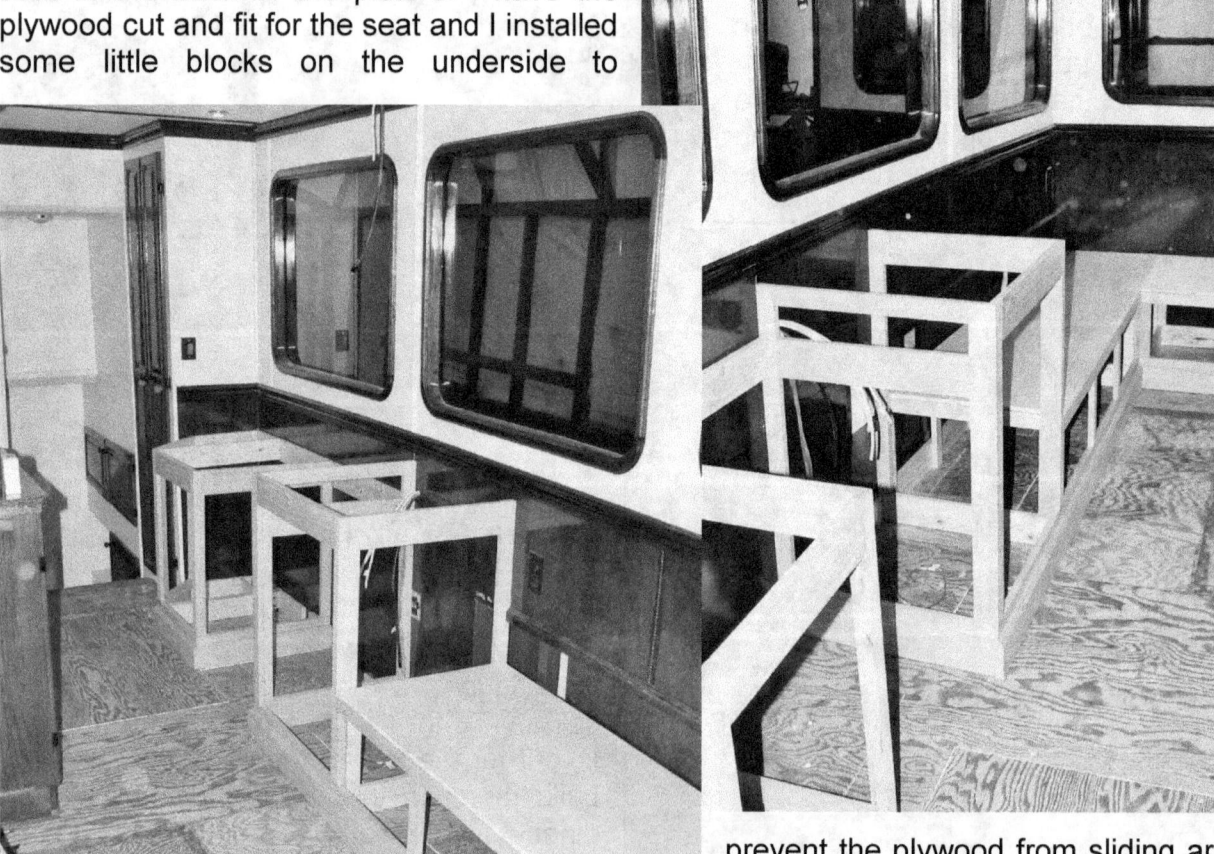

prevent the plywood from sliding around on the frame. To the left is a view looking forward and shows the entertainment center and the desk a little better. I spent the day today working on the desk getting it lined up with the entertainment center and installing some of the drawer partitions.

Saturday, March 16, 2013

I've been working hard on my cabinets in the saloon. As you can see in the picture to the right and below, I have them all filled, sanded, and stained, ready for the first coat of varnish. I've also been working on my desk top so hopefully it will be ready to install once I get the varnish

completed. The picture above is looking aft toward the dinette. You can see the bench seat for the dinette which will convert into a bunk for guests. I still have to build the table. The picture to the left is looking forward and shows the navigation station desk and the entertainment center cabinet a little better.

Sunday, March 24, 2013

Since I have my cabinets all stained and ready for varnish, I've been working on my drawers and the desk top. To the right is a picture of some of my drawer fronts, desk top, and my cutting board that goes under the galley cabinets. I had to order a piece of maple for the cutting board because I couldn't find it anywhere around here. I used a couple pieces of African mahogany for the pulls on each end of the cutting board. I have them glued and clamped in the picture to the right.

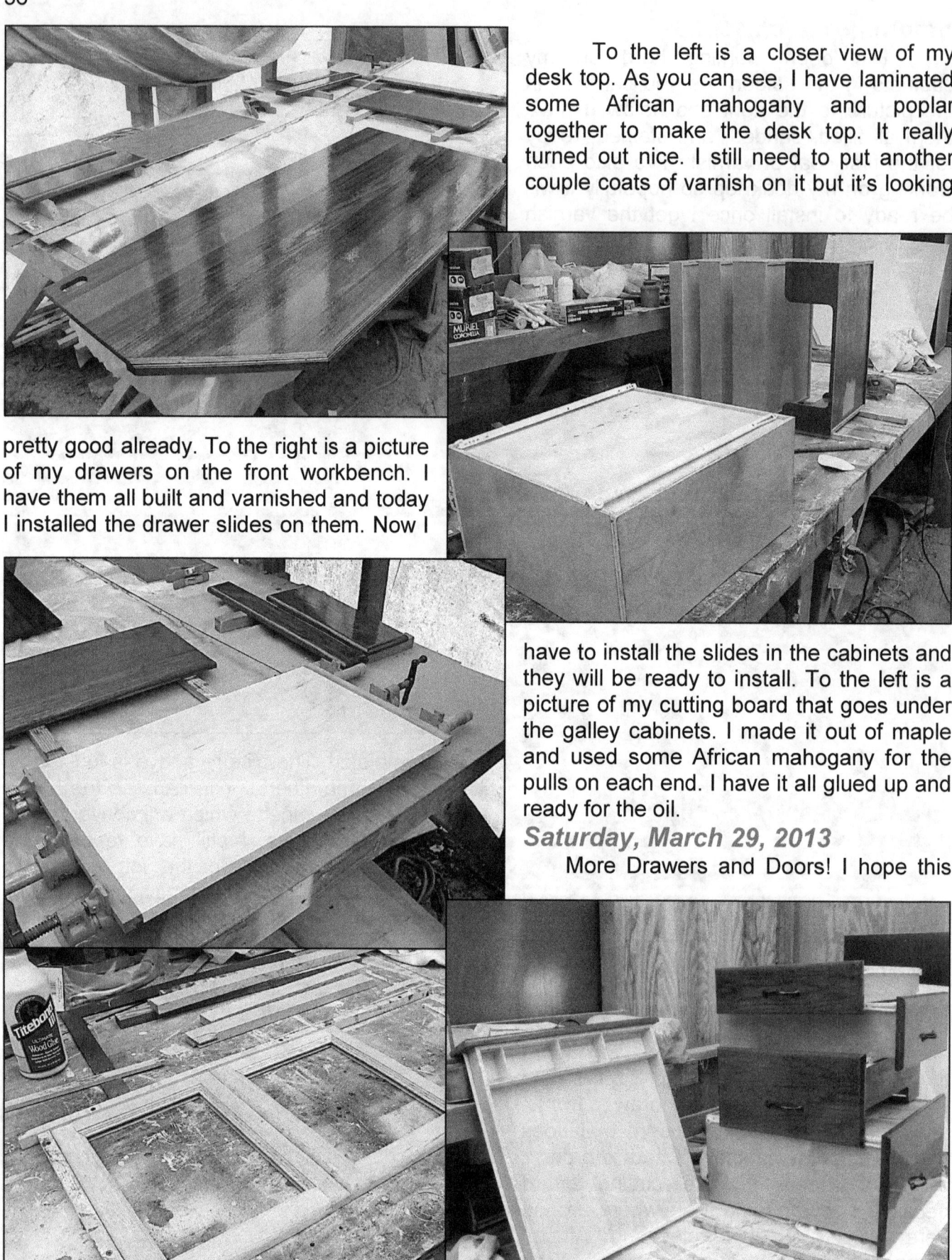

To the left is a closer view of my desk top. As you can see, I have laminated some African mahogany and poplar together to make the desk top. It really turned out nice. I still need to put another couple coats of varnish on it but it's looking pretty good already. To the right is a picture of my drawers on the front workbench. I have them all built and varnished and today I installed the drawer slides on them. Now I have to install the slides in the cabinets and they will be ready to install. To the left is a picture of my cutting board that goes under the galley cabinets. I made it out of maple and used some African mahogany for the pulls on each end. I have it all glued up and ready for the oil.

Saturday, March 29, 2013

More Drawers and Doors! I hope this will be the last of them! The picture above

Saturday, March 16, 2013

I've been working hard on my cabinets in the saloon. As you can see in the picture to the right and below, I have them all filled, sanded, and stained, ready for the first coat of varnish. I've also been working on my desk top so hopefully it will be ready to install once I get the varnish

completed. The picture above is looking aft toward the dinette. You can see the bench seat for the dinette which will convert into a bunk for guests. I still have to build the table. The picture to the left is looking forward and shows the navigation station desk and the entertainment center cabinet a little better.

Sunday, March 24, 2013

Since I have my cabinets all stained and ready for varnish, I've been working on my drawers and the desk top. To the right is a picture of some of my drawer fronts, desk top, and my cutting board that goes under the galley cabinets. I had to order a piece of maple for the cutting board because I couldn't find it anywhere around here. I used a couple pieces of African mahogany for the pulls on each end of the cutting board. I have them glued and clamped in the picture to the right.

To the left is a closer view of my desk top. As you can see, I have laminated some African mahogany and poplar together to make the desk top. It really turned out nice. I still need to put another couple coats of varnish on it but it's looking pretty good already. To the right is a picture of my drawers on the front workbench. I have them all built and varnished and today I installed the drawer slides on them. Now I have to install the slides in the cabinets and they will be ready to install. To the left is a picture of my cutting board that goes under the galley cabinets. I made it out of maple and used some African mahogany for the pulls on each end. I have it all glued up and ready for the oil.

Saturday, March 29, 2013
More Drawers and Doors! I hope this will be the last of them! The picture above

shows all my drawers finished and ready for installation. The picture to the left on the previous page shows the door frames for my entertainment center. I'm going to put dark glass in these doors because most remote control devices use infrared to transmit information to the peripheral device. I thought I would try to etch "Molly B" into the glass just for a touch of class. I've never tried etching before so it might be interesting.

Tuesday, April 02, 2013

I have finished my cabinets in the saloon except for the top of the entertainment center which is all stained and ready for varnish. I ran out of varnish so I guess I will have to spring for

another can so I can finish up. In the meantime, I've been working on my dinette table. I have it all built and put together except for the top of the table. I need to get that stained and put my fiddle molding on and then get the whole thing varnished and it will be done. To the left is a picture of my table with the pedestals attached to the bottom of the table. The two pedestals in the foreground of the picture are the ones that will be secured to the deck. The two long narrow pieces are the removable part of the pedestals which slide into the short ones which will allow the table to be dropped down and convert into a bunk. I just need to turn the table over and install my fiddle molding, stain the top, and it will be ready for varnish. Then I can install it and see if it will work as planned. I never seem to know for sure about these things until I get everything done.

Thursday, April 04, 2013

I've completed my table finally. At least it's all built. I still need to put the varnish on it and then it will be all done. To the left is a picture of my table top with the fiddle molding installed and all stained.

To the right is a little closer view of my table. You can see the fiddle molding a bit better in this picture. It turned out pretty well except that I have a few scratches in the top so I think I may have to do a little more sanding and staining before it's really all done.

Saturday, April 06, 2013

I finally got my air horn installed and working today. I had it working before but I had a small air leak at the tank and it took me most of the morning to get it fixed. To the left is a picture of my little compressor and the air tank which I mounted under the cabinets on the bridge. The pressure switch shuts the compressor off at 110 PSI and when you hit the horn button it really sends out a blast at 139 decibels. I have three horns in a cluster which really sound nice. To the right is a picture of my horns. It looks

like there are only two horns in this picture but there are actually three. I just didn't get the right angle for the picture to show all three of them.

Tuesday, April 16, 2013

I had to have a hernia repaired last Friday, so I'm on the mend again. But, I'm able to work at least a few hours a day so I'm making a little progress anyway. To the left is a picture of my handrail to the bridge. I almost have it all put together and ready for stain and varnish. I can't make up my mind whether I want to install it and then finish it or if it would be better to finish it and then install it. I'm leaning toward finishing and then installing as this way I don't have to worry so much about getting stain and varnish on places it doesn't belong.

The spindles in this view appear to be slightly slanted; however they are all plumb so they are really straight up and down. It might just be the camera angle that makes them look this way.

To the right is a view of the access panel below the stairs. (At the bottom of the picture) This panel must remain removable to allow access to the compartment below the stairs where all the wiring goes from the bridge to the engine room. I have notched the panel for each of the spindles so it will still be removable and I can access the compartment.

Tuesday, April 23, 2013

I finally have my cabinets all put together. I only have the galley rail to install on the entertainment center as you can see in the picture below, and the cabinets will be all done! In the view below you can see the entertainment center and my desk. This view is looking forward toward the bridge.

You can also see part of my dinette table in this view. I got a really nice finish on the table top as well as on the desk and entertainment center tops. You can see the shine in this view of all three of them. It took about six coats of varnish to get the grain of the oak filled so I got a nice solid finish. I had to do a lot of sanding between coats as well. You don't have to sand much, just take the shiny off the varnish and that levels the finish out and eventually gives you a real nice gloss without any indentations, or at least very few of them. You occasionally have a spot that you just can't seem to get filled and you have to live with it but most of the time you can get all the low spots filled so you end up with a real nice smooth finish.

To the left is a view looking aft and shows my dinette seat and table a bit better. I need to get the patterns made for the cushions for the dinette seat and this end will be done as well. I'm going to wait to mount the table pedestals to the deck until I have the flooring all installed and then screw the pedestals down to the deck on top of the flooring. That way I don't have to try and cut the flooring out around the base of the pedestals.

Below is a view of my dinette seat and table. You can see the double pedestals in this view much better. It's a

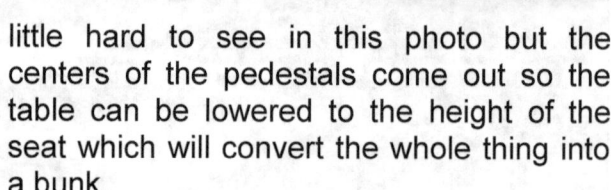

little hard to see in this photo but the centers of the pedestals come out so the table can be lowered to the height of the seat which will convert the whole thing into a bunk.

To the left is a little closer view of my navigation station, or my desk. It's a little dusty on the top but I think you can get an idea of the nice shine on the top. I was doing some sanding on my stair treads in the saloon when I noticed the dust so I moved my sanding operation to the aft deck.

Saturday, April 27, 2013

Today I installed the doors on my entertainment center. To the right is a picture of the installed doors. You can see my glass etching in this view. I etched the glass and then

used a paint stick to whiten the etching to make it stand out a little better. It turned out really well.

To the left is a picture of my handrail going up into the bridge. I finally got all my stair treads varnished and installed so I could finish the handrail. As you can see in this view, I added a post at the bottom of the railing to help stabilize it laterally. The spindles are just too skinny and they bend a bit when you push on them from the side.

I have also been working on the stairs going down to the forward cabin. There are two steps that have to be removable to access the house bank battery box which is under these steps.

Here are a couple pictures, above and to the left, of the two steps that will be removable. Actually they really aren't removable, just hinged so they can be lifted out of the way to access the battery box. You can see the piano hinge in the view to the left.

Sunday, May 05, 2013

I have finally completed the stairs and handrail down into the forward cabin. To the right is a picture of the two steps that are hinged so I can access the house bank of batteries. It's a little hard to see but there is a latch about midway up the right side of the picture that holds the two steps up so I can access the batteries.

To the left is a view of the stairs with the two hinged steps in the closed position. You can see the latch a little better in this view as well as the handrail. Now I have all my stairs and handrails completed and I can move on to the next item on my loose ends list. There are a lot of items left on the list, even though I have already cleaned up over two pages of them.

Below is a picture of my entertainment center. I have the galley rail all done and ready to install on the top of it

as you can see in this picture. This is just one more item to cross off my list. I also spent some time today putting the hinges on the two doors on the anchor chain locker. I finally got those doors hung. All I need to do now is figure out what kind of a latch to put on them to keep them closed. Next I think I will work on installing the flooring on the bridge since I have it all put back together again. I finished installing the labels and wiring in the overhead console so the next thing is the flooring.

Wednesday, May 08, 2013

I've been working on my TV cabinet for the past few days now and I pretty well have it all built. To the left is a picture of the top portion of the cabinet. This is the part that gets secured to the ceiling. The TV will be mounted into the other part of the cabinet which I don't have a picture of, which will be hinged to this part and swing down for viewing. It doesn't look like much in this picture, but it was really a challenge to build. I hope the TV fits into it, I haven't tried it yet!

I've also been working on the flooring on the bridge. To the left is a view of the bridge sole and the flooring installation. As you can see, I have put a coat of primer on the sole to seal it before I started installing the flooring. Below is a view of the flooring installed. I still need to

install the quarter round molding around the edges of the floor before it will be completely done, but I have the hard part done!, I think!

Saturday, May 11, 2013

I have my TV cabinet all put together and painted and it's ready to be installed in the saloon. I think I'm going to need some help getting this thing held up to the overhead while I screw it in place. To the right on the previous page is a picture of my deck in the saloon. I have it all sanded and primed, ready for the flooring. To the left is another view of the floor all primed and ready for the flooring.

Tuesday, May 14, 2013

I've made some pretty good progress with my flooring in the saloon. It's amazing how close the flooring matches the cabinets. It's a little hard to tell with the black and white pictures, but it really does match very well. You can see a little bit of the flooring in these two pictures, in the galley and also around the dinette area. I have it all completed now with the exception of the quarter round molding. I have some of that installed but I ran out so I had to stain and varnish some more to finish up.

I have also been working on my TV cabinet. I have it mounted to the overhead and ready for the installation of the TV. I won't be putting the TV in the cabinet until we are ready to launch because I'm going to use the same TV we are using at home now in the boat.

To the left is a picture of my cabinet installed on the overhead. I still need to put some trim around it but otherwise it is done.

To the right is a view of the cabinet opened up and shows where the TV will be mounted. I still need to figure out how to hold the cabinet in the open position so it won't swing back and forth. I think I might try to install a hydraulic stay, or maybe just a regular window stay to hold it in place. I have a latch on the front of the cabinet to keep it closed, I just need to be able to keep it open and I will be all set. I think my HDMI cable is a little short as well, so I will have to see if I can find and extension so it will reach where it plugs into the back of the TV.

I have an air leak in my air horn system, and I've been unable to find it so I called the people I purchased it from and they said to send the compressor back to them for testing and they will replace it if it is bad. I have checked every connection several times with soapy water and can't find any leaks so the only place left is the compressor itself unless the tank is leaking somewhere. There is no check valve in the line from the compressor to the storage tank, so the pressure of the tank is being applied back to the compressor itself. Maybe the pressure is leaking past the rings or valves inside the compressor. I just don't know so I'm sending it back for them to test it. I sure hope they can find the problem, because I sure don't want that compressor coming on every few minutes, it makes too much noise.

Tuesday, May 21, 2013

I spent the day today working in the saloon. I thought I was done in there but I found a few more things I had to finish up. My dinette table needed to be relocated just a bit because it was a little crooked. When I attempted to relocate it I hit a screw under one of the pads that gets screwed down to the deck. The odds of doing that are astronomical but I did it. So, I had to find a new location for the table which I finally completed. I also needed to finish painting the area under the seat of the dinette, and then I had to put my kick plates on the back door to cover the hole in the door. Now I can get back to work in my engine room. I needed a break from that because I'm a little sore from crawling around down there for the past few days. I have managed to get some of my engines cleaned up and repainted, and I have also completed the installation of my emergency bilge pump system. I'll have to get some pictures of that set-up, it's kind of neat.

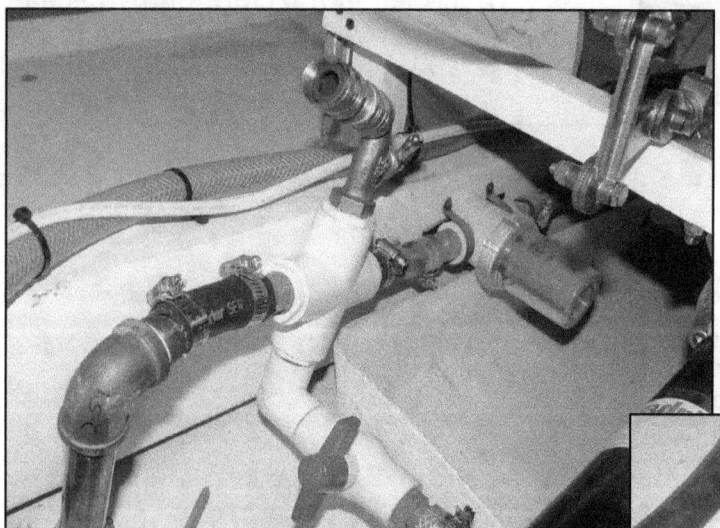

Wednesday, May 22, 2013

I got a couple pictures of my emergency bilge pump systems today. To the left is a view of the port set-up. The Seacock is in the lower left corner of the picture. I installed a 4-way so I could install a valve on top to attach a hose to so I could run the engines on the hardstand. I will just close the seacock and open the hose valve. Then below that I installed another valve that is attached to a hose that runs to the lowest part of the bilges where if the regular bilge pump can't keep up with the inflow of water, I can close the seacock again, and open the bilge hose valve and the engine driven raw water pump will pump the bilge water overboard via the exhaust hose while cooling the engine.

To the right is a view of the set-up on the starboard engine. It is basically the same except that the suction hose running to the bilges is facing forward instead of amidships as it is on the port side. This is because the engine stringer it is attached to is taller on the inboard side of the engine. With both engines pumping bilge water

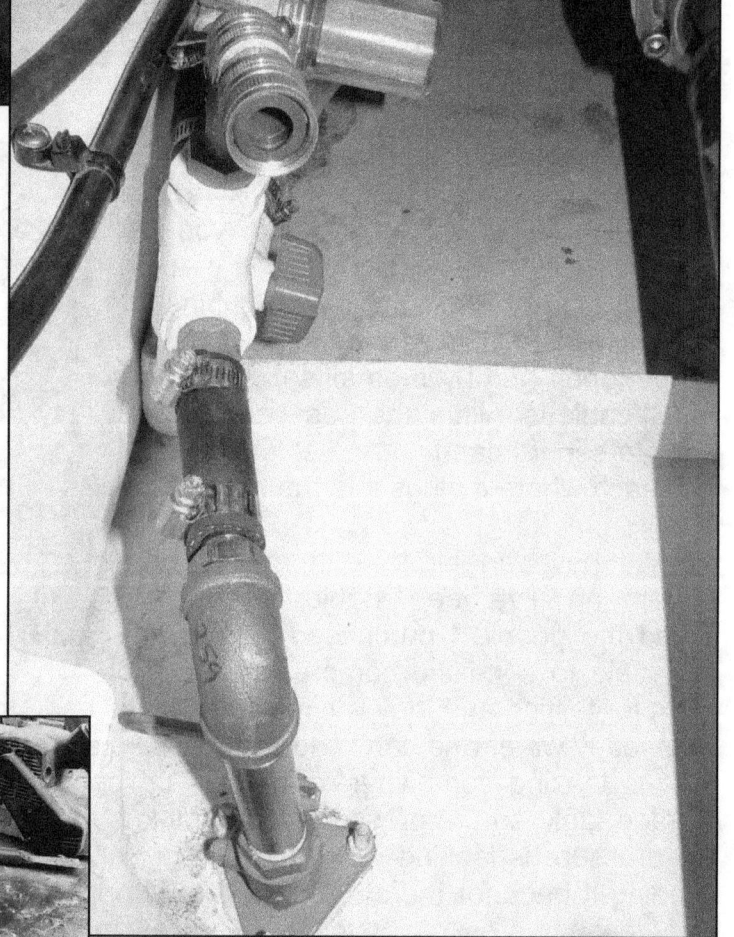

overboard along with the regular bilge pumps, we should be able to move a lot of water and get the boat to shore before she goes down.

Thursday, May 23, 2013

I spent most of today making a fairing block for my depth sounder transducer. To the left is a picture of the block I used to make the fairing. The location I chose for the transducer had a 10 degree rise of the hull athwart ship and a 2 degree rise astern. So I had to make the fairing with these tapers to allow the transducer face to be held parallel to the surface of the water. The block shown in the picture on the previous page shows the 10 degree taper. To the left is a little closer view of the block showing the 10 degree taper. The 2 degree taper runs the length of the block so I had to plane one end down about ¼ inch to make the 2 degree taper. The picture to the right shows the block cut down to the basic shape of the fairing, with the hole cut in it for the transducer. Below is a picture of the finished fairing block with the transducer inserted. The shape of the fairing block will allow the water to smoothly transition from front to back over the transducer with a minimum of disturbance.

Sunday, May 26, 2013
I got a couple pictures of my fairing block installed today. The picture below

shows the fairing block installed with the transducer face parallel with the world. The 10 & 2 degree corrections in the fairing block worked out just right. As you can see, I tapered the block fore and aft to provide a little better turbulence reduction while underway even though the Molly B is not really a high speed vessel; I still want the least amount of drag from the hull as possible. Every little bit helps the end result when it comes to fuel consumption.

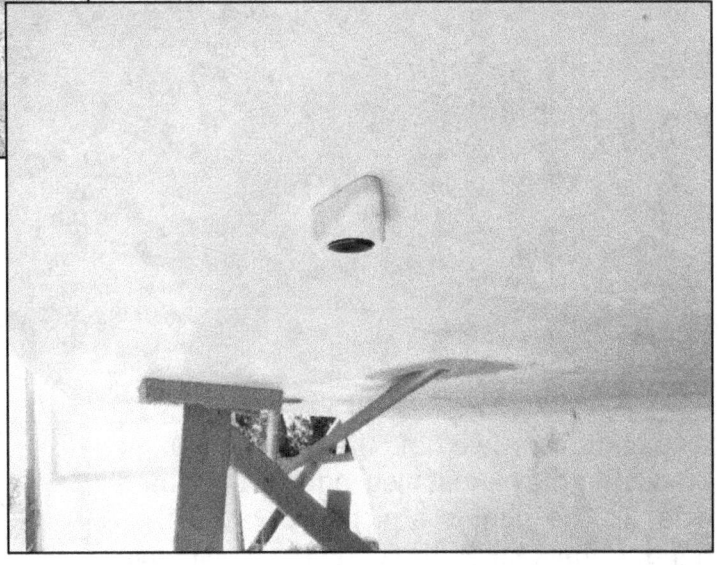

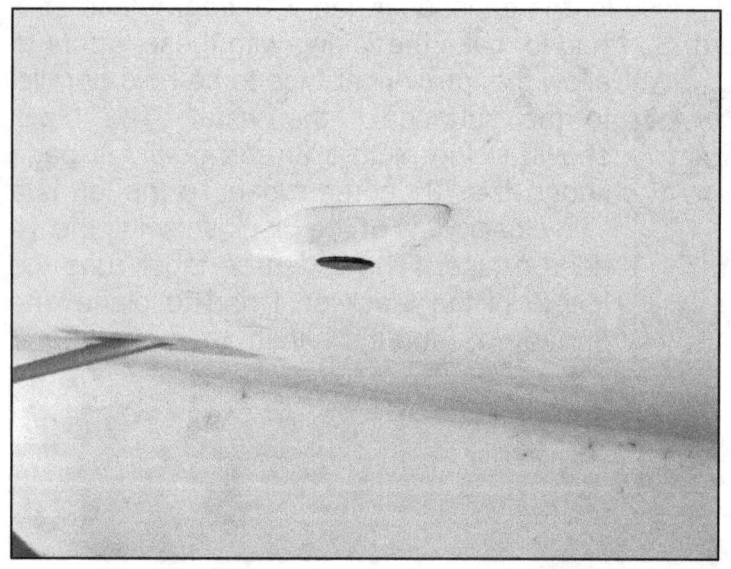

To the left is another view of the fairing block from a little different angle. You can see the tapers on both ends of the block better in this view. The black circle in the middle of the block is the face of the transducer. I still need to screw the block to the hull, but I'm waiting for my 3M marine sealant to come in before I do that so I can get the whole thing sealed really well. I sure don't want any leaks around this thing.

Sunday, June 02, 2013

I'm still working down in the bilges and it looks like I'll be there for a while yet. I've been working on my engine exhaust hose hanger brackets. I need to make sure the hoses are sloping downhill to the transom so the water in the exhaust runs out the back of the boat. That means each of the hanger bracket has to be a different length. I have three brackets installed on the starboard side and three on the port side. In the picture below, I have the last three brackets made and ready to install and then I think I should have the hoses pretty well supported. I still need to install the exhaust hose for the generator which may need one or more hangers to support it. I'm waiting for my wet elbow which has been in the shop under construction for the past year and a half. I told the welding shop to open it up and I would try to fix the leak in it myself. Once I get that done I can install the hose for the generator. Below is

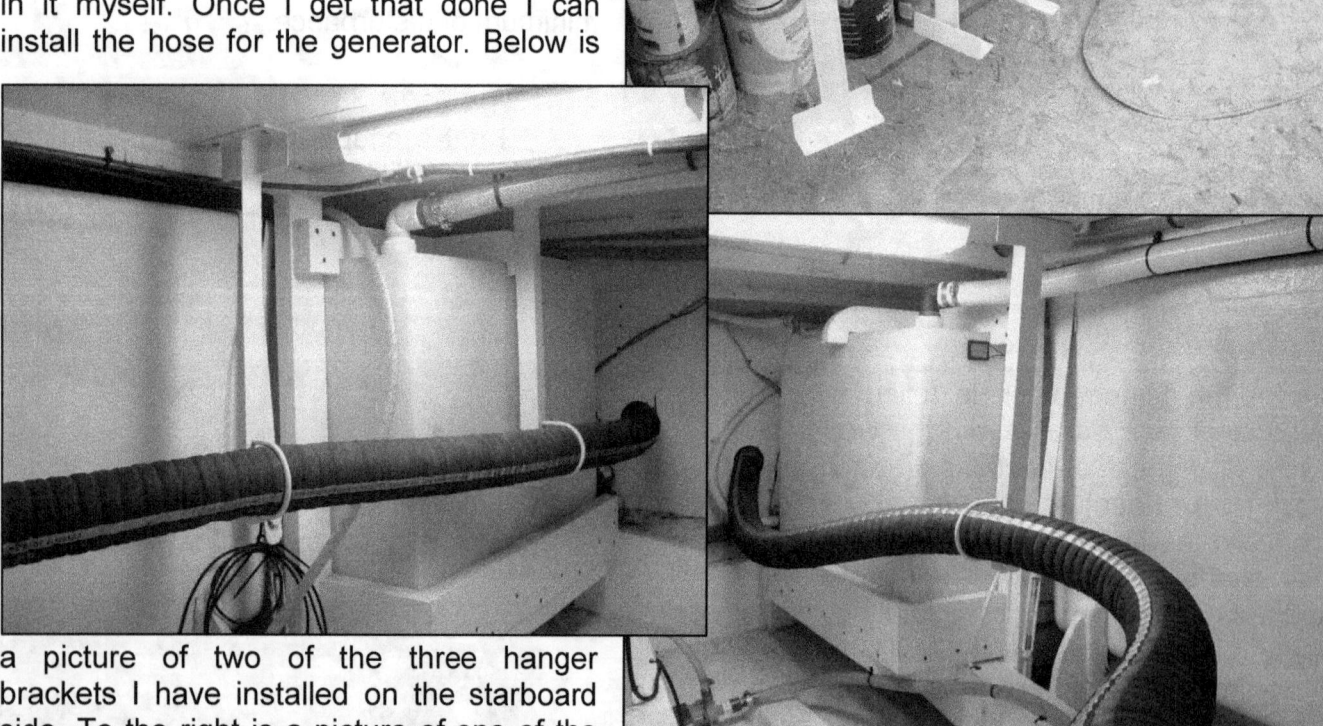

a picture of two of the three hanger brackets I have installed on the starboard side. To the right is a picture of one of the hanger brackets on the port side. I still have two more brackets to install on this side of the boat.

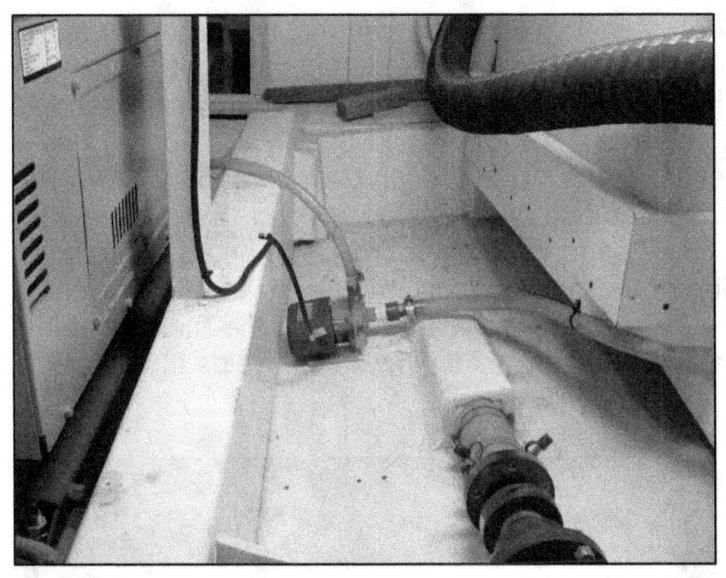

I have completed the installation of the water pump for the generator exhaust. I wired the pump directly into the 110 VAC outlets on the generator so as soon as the generator starts and reaches the proper rpm, the pump will start pumping water to the wet elbow. The picture to the left shows that little pump installed alongside the generator. I picked up my wet elbow from the welding shop yesterday, and I brazed over the welds inside the elbow which sealed all the leaks. So, now I just have to take it back to them so they can weld the top back on and then I can measure and cut the exhaust hose for the generator. Then I can tell whether or not I might need any additional hanger brackets for that hose.

Wednesday, June 12, 2013

To the right is a picture of what I call my wet elbow. I've been waiting for this thing for the past year and a half. Getting it fabricated has been a real challenge. The welding shop that has been working on it

has had their fair share of problems with this thing. When they first made the thing, it would leak water between the upper compartment and the lower compartment. As you can see in the picture above, the water goes into the upper compartment via the bung on top of the elbow, and the exhaust comes into the lower compartment from the generator. In the picture to the left, you can see the baffle between the two compartments. The exhaust hits the baffle from below and the water hits it from above thus cooling the exhaust and then both are discharged out the exhaust hose through the transom. If water leaks into the lower compartment, it will run into the muffler of the generator and possibly into the engine itself. So, they tried to fix the leak for months and just couldn't get it to stop leaking. I finally asked them if they could cut the top of the thing off, and I would take it to my shop and sweat some brass into the cracks in the welds or pinholes or whatever was causing the leak. They did this for me, and I took it home and brazed all the welds inside the box around the baffle. I checked it for leaks, and there were none. I had it fixed! I took it back to the welding shop and asked them to weld the top back on and they did. The only problem was that they couldn't get the top compartment to stop leaking. Every time they welded it, it would develop more cracks and continue to leak. So, I brought it home again and I have ground down all the welds on the outside of the box and it is now ready to be brazed and hopefully sealed to stop the leaks. You can actually see several cracks in the stainless steel where they have welded it so I hope I will be able to fill them like I did with the welds on the inside of the box around the baffle.

Thursday, June 13, 2013

I've spent most of the morning today working on my wet elbow. To the right is a picture of it with the brass put on it. It looks pretty bad, but it doesn't leak! I'm letting it cool and once cool I will sand down the brass and clean it up a bit. It will look a little better, but will still look pretty rough. I guess I shouldn't worry so much about what it looks like as long as it functions as it is supposed to. I got it all cleaned up and cut some slits into the exhaust inlet pipe so I can squeeze it tight around the muffler pipe with a clamp. I installed in on the generator and it fits pretty well. I measured and cut

the exhaust hose to length. I should be able to get it installed tomorrow.

Friday, June 14, 2013

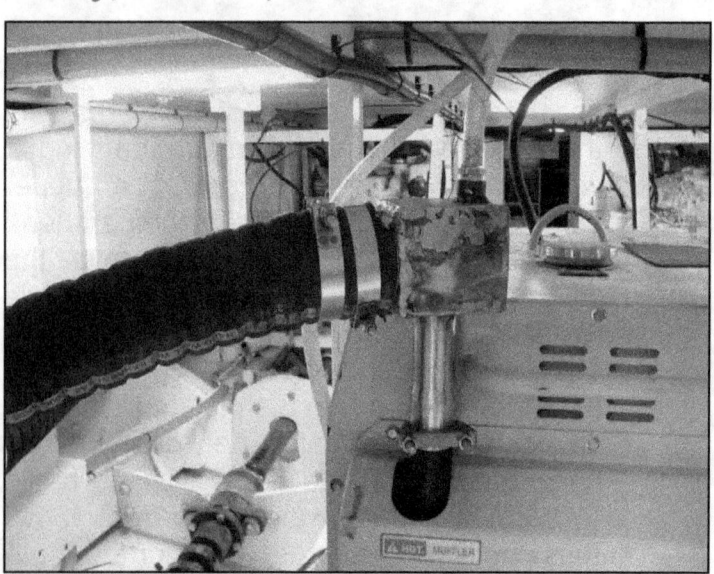

I leak tested my elbow one more time, and it seems to be good, so I installed it on my generator. The picture to the left shows the installation. I have the exhaust hose connected to the wet elbow, but I still have to drill the hole in the transom for the flapper valve and then I can finish installing the exhaust hose. I have the port engine exhaust hose installation completed, and I have just finished installing the mounting straps for my fire suppression system. I'm making some pretty good progress and it won't be long and I will be ready to launch.

Sunday, June 16, 2013

I completed the installation of my flapper through the transom and installed the exhaust hose from the generator to the flapper. I needed a support bracket for the port engine exhaust hose to keep it sloping toward the transom so I got that made and installed. I made the bracket long enough to support the generator exhaust hose as well. Then I added a few wire ties to hold the two hoses together and on the support bracket. It worked out well and I have the two hoses completely installed now. To the right is a picture of where the two hoses go through the aft bulkhead into the steering locker. You can see the support bracket under the two hoses on the

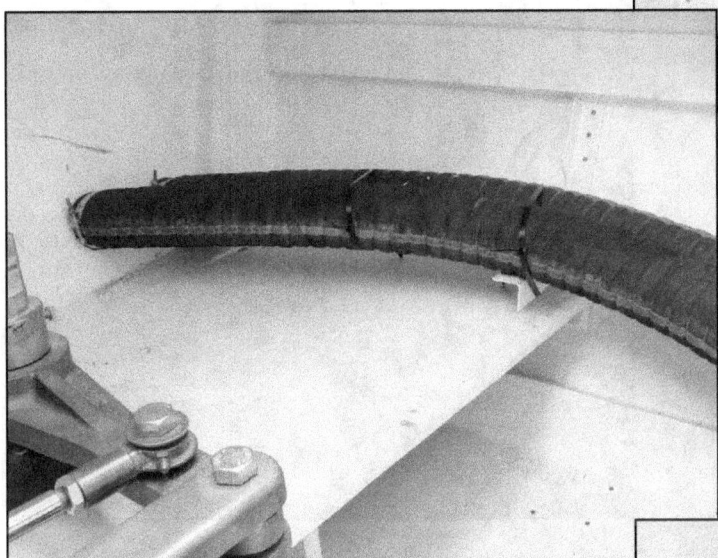

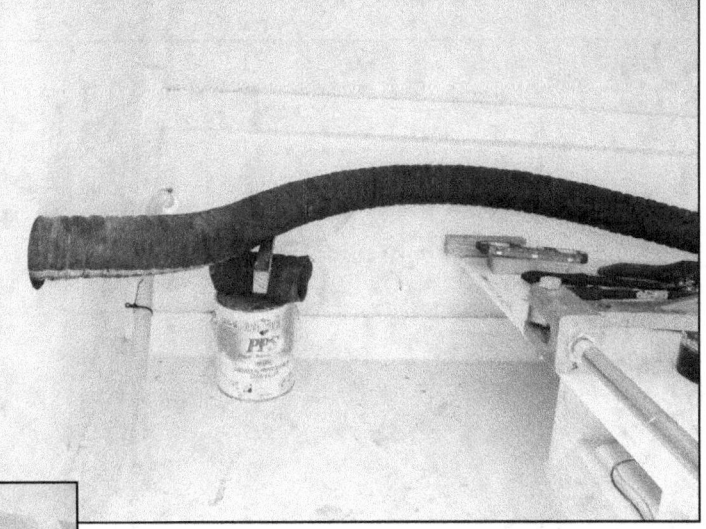

right side of the picture to the left, just above the rudder shelf. As you can see, the two hoses are tied together with wire ties to keep them lined up properly, and attached to the support bracket.

To the right is a picture of the starboard exhaust hose where it goes through the aft bulkhead into the steering locker. I have the hose propped up with a paint can and some other things to level it and get a measurement for the support bracket I will need to install there to hold the hose up. I didn't need a support bracket here on the port side because the hose came through the bulkhead a bit lower on

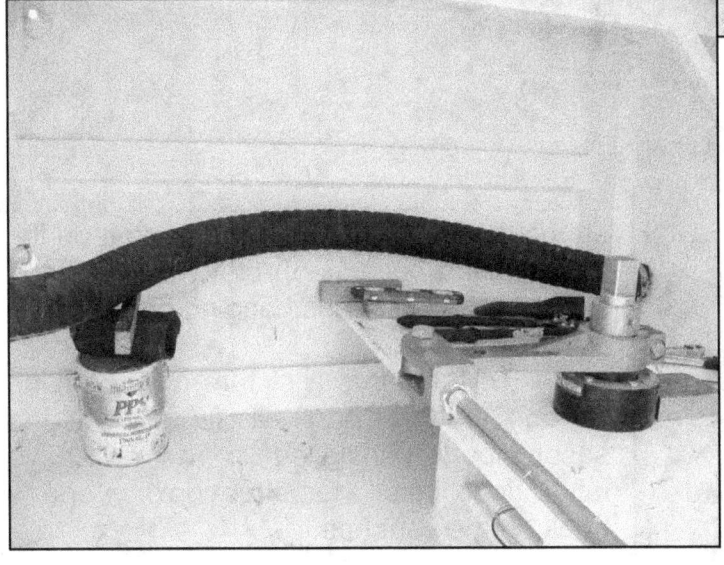

that side and didn't need support there to keep it sloping properly. I will also need a support bracket a little farther aft to hold the hose down just a bit so it will slop properly toward the transom. As you can see in the picture to the left, the hose needs to be pulled down just a bit where it goes over the rudder shelf to keep it sloped properly. I don't have enough angle iron left so I will have to get another short piece to finish these brackets. It won't be long and I will be done with these hoses.

Wednesday, June 19, 2013

Here is a picture of the port engine exhaust hose installation completed. You can see all the hanger brackets I used to hold the hose in the proper position. To the

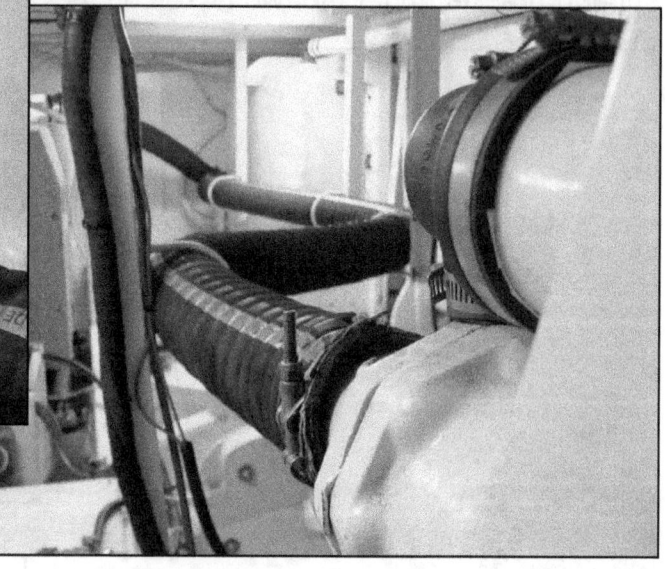

right is a little different view of the port exhaust hose. You can see a few different hanger brackets in this view. Below is a view of the starboard exhaust hose. This side didn't have so many bends in it, and went pretty much straight back to the aft bulkhead. Below is a view of my fire

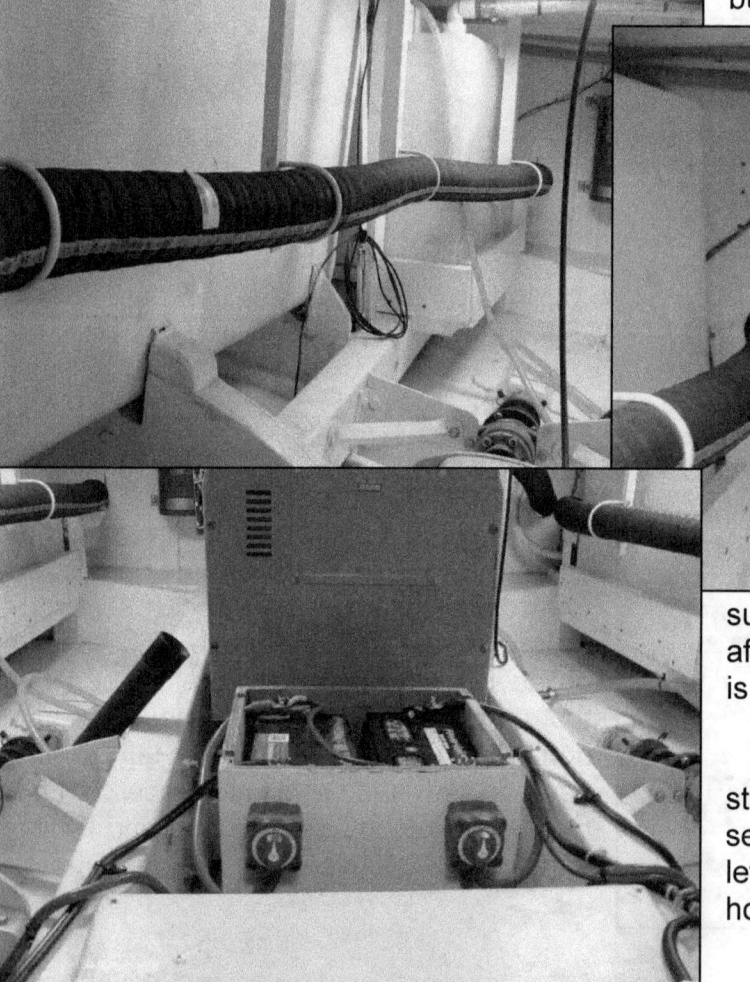

suppression cylinder. I had to hang it on the aft bulkhead because the forward bulkhead is already full of things hanging on it.

To the left is a picture of my engine starting battery bank. I finally got the second battery installed so now all I have left to do is install the battery box cover and hook up the vent hose.

Monday, June 24, 2013

I've spent the past few days working on the engine shutdown system for my fire suppression system. To the left is a picture of the control circuit board with a few of the wires connected. I still have a few more wires to hook up and it will be done. This is the control center for the shutdown system. There is a pressure switch on the cylinder down in the engine room, and that is wired into the circuit board so when the cylinder discharges, it will send a signal to the circuit board and activate all the solenoids that shut down the engines, generator, and the engine room blowers. This system is required in a boat that has diesel engines because the agent in the cylinder will not necessarily shut down a diesel engine like it would a gas engine. If the diesel keeps running, it will suck the agent out of the engine room and prevent it from extinguishing the fire. As a result, both ABYC and the Coast Guard require this system be installed in a diesel powered vessel that has a fire suppression system installed.

The system also has a manual discharge system that allows you to discharge the cylinder manually. To the right is a picture of the "T" handle that activates the discharge of the cylinder manually.

There is also an override switch on the system monitoring display that allows you to override the shutdown system and restart the engines if you are in a dangerous situation. It is quite a complicated system when it's all put together, and it's not exactly easy to set up if you aren't a marine electrician but it is

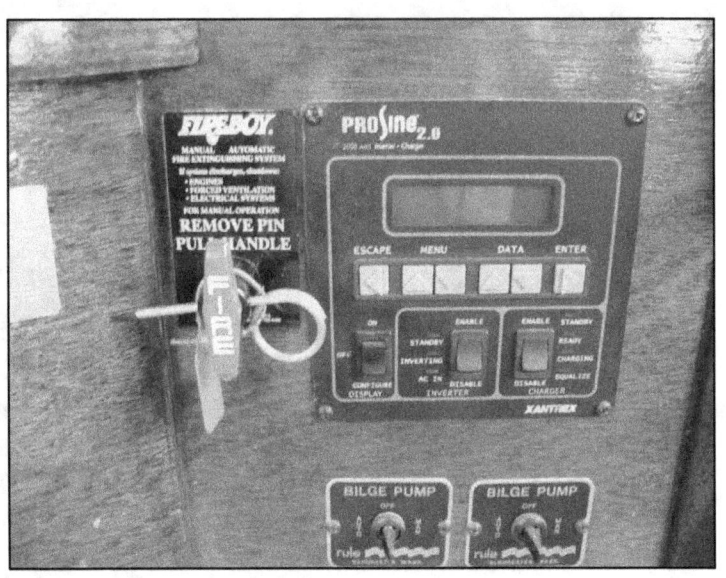

doable if you understand the wiring of your vessel. The system comes with a pretty comprehensive schematic so it's pretty easy to figure it out.

Saturday, June 29, 2013

I ran into a problem installing my fire suppression system. I can't seem to figure out how to hook up my generator so I decided to take a break and work on my radar mount. I purchased a Scanstrut mast mount and the mounts feet are designed to be attached to a round metal mast. So, I had to make some new mount feet to fit my

square wooden mast. The picture to the left on the previous page shows one of my mount feet after it was cut out of a 2 x 2 x 1/4 inch aluminum angle. This ended up being a bigger project than I anticipated because the mounting holes on the mount I purchased were almost even with the sides of my mast. As a result, I had to make the feet extend out farther than the side of the mast in order to have room to install the bolt which holds the feet to the mount. Below are a

couple pictures of the mount feet all cleaned up along with the spacers ready to be assembled onto the radar mount. As you can see in the pictures to the left and right below, there is a gap between the feet and the side of the mast. That's why I needed to make some shims to fill the gap. The gap resulted from the radar mount being just a little too narrow for my mast as I mentioned before. Had I known this, I guess I could have made the mast just a little narrower. But, with the spacers the mount fits my mast just right. I cleaned all the feet and spacers and gave them all a coat of paint to help keep them from oxidizing.

To the left is a picture of all my mount feet with a coat of primer on them hanging on the bench to dry. I gave them a coat of white once the primer was dry, and now they are ready to assemble to the radar mount. Now I just have to figure out exactly where on the mast I need to mount the whole assembly. I need to mount the radar dome as high as possible on the mast but I need enough room above the dome to mount my yard arm on which I will mount my deck lights, TV and GPS antennas. I will also have to leave enough room above that to mount my masthead light and anchor light. My VHF antenna I think I will mount on the back side of the mast above the radar dome. I need to keep that antenna out of the radar beam so as not to cause any interference or conflicts between the two antennas.

Saturday, July 06, 2013

I finally figured out what the problem was with my engine shutdown system. The circuit board/control box I purchased is a three circuit box, and I can't hook up four systems to a 3 circuit box. The factory recommended the three circuit box and when I called them to ask how to hook up my generator along with everything else, they said I couldn't do it and I had to upgrade to the five circuit box. I sent the three circuit box back to them and they upgraded me to the five circuit control box. I got it yesterday, and spent most of the day today installing it.

This control box is quite a bit more complicated than the three circuit box shown on page 53. You can see in the picture above that it has a few more terminals to contend with. I labeled all my wires when I removed the other control box so I would be sure to get the right wires on the right terminals when I got the new control box. I have all my wires connected except my generator which will be attached to one of the three auxiliary terminals at the bottom of the box. I thought I had the correct wire isolated at the generator ignition switch but it turned out to be the wrong one so now I have to go back and try to find the right wire to connect to the shutdown system.

Thursday, July 25, 2013

I've been pretty busy since I got back from visiting the kids in Virginia last week. Hadn't seen them or my grandchildren for over two years so I thought it was about time I took some time off from the Molly B and took a trip to see them all.

Yesterday I designed a mount for my anchor light and got the light attached to it. Now I can mount my light as soon as I get the mast painted. I will have to get a picture of it tomorrow. I also got a call from Randy at the upholstery shop and he said he had my cushions done for the dinette and the bench seat on the bridge. Above and to the left are a couple pictures of the cushions in the dinette. They are a burgundy color and really look nice. I haven't installed the ones on the bridge yet because I have the bench seat up there full of tools and my ships wheel and some panels. I need to get it all cleaned off before I can install the cushions up there.

I've also designed a swivel for my mast boom. Below is a drawing of the assembly. I'm having it made at a welding shop out of stainless steel. I hope it works the way I think it will. If it does, I will be able to swivel the mast up and down, as well as left to right. This will enable me to offload my dingy as well as lower the mast if I need the extra clearance passing under a bridge. I have also designed a hinged mast step which will allow me to drop the mast by raising the boom all the way up and then using the forestays I will be able to drop the mast to the boat deck and then raise it back up again when I'm clear of the obstruction.

The welding shop completed my two items but when I got there to pick them up I found they weren't made just right so I had to leave them there for them to fix. I should be able to pick them up tomorrow. I have a picture of a drawing of the hinged mast step which shows how it will work on the next page. There is a hole through the forward portion of the step which will lock the mast in the vertical position.

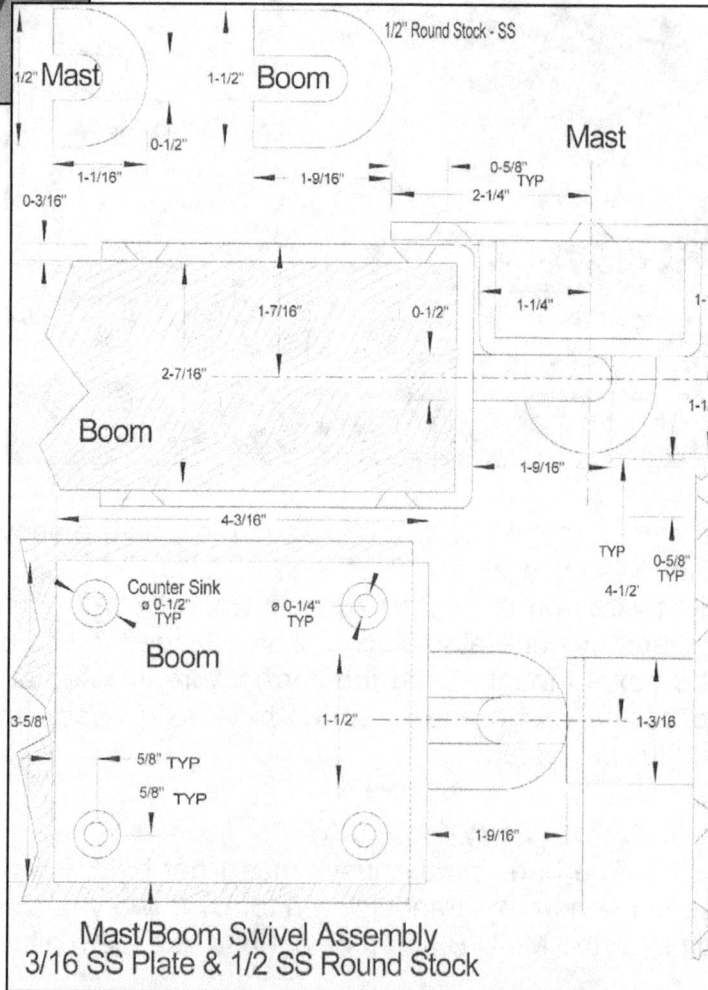

Mast/Boom Swivel Assembly
3/16 SS Plate & 1/2 SS Round Stock

Mast Step Hinge

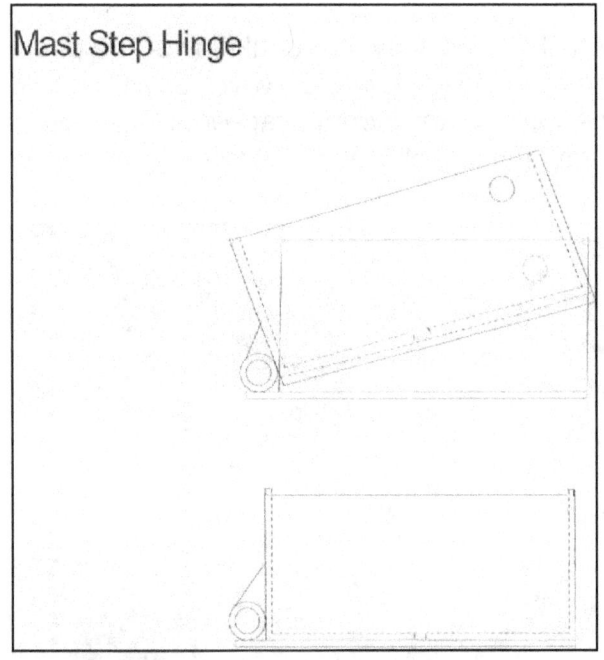

To the left is a picture of a simple drawing showing basically how the step hinge will work. You can see the hole in the top drawing that will lock the mast in the upright position. As soon as I get them back from the welding shop, I will get some pictures to show you what they look like.

Friday, July 26, 2013

I made another trip to the welding shop today. They had my mast boom swivel completed, so I picked it up but the mast step hinge wasn't finished yet so I will have to go back up on Monday to pick it up. I think my design for the mast boom swivel is going to work, however I need to have the welding shop add a small shim to the inside of the boom ring so it doesn't drop down and lock the boom in that position. Below is a picture of the mast boom swivel.

The flat plate is screwed onto the mast, and the smaller channel is screwed onto the end of the boom. As you can see in the picture to the right, the boom would be swiveled to the starboard. The design works great swiveling port to starboard but when you raise and lower the boom there is a small problem. In the picture below shows

the boom raised all the way up. This would be the required position of the boom when I need to lower the mast. The problem is the gap in the loop on the end of the channel. When the boom is raised, the loop slides down and locks the boom in the up position. The picture to the right shows the position of the loop on the boom bracket with the boom in the up position. You can see a gap in the loop on the boom bracket which causes the boom to be locked in the up position. The only way you could lower the boom would be to push up on the boom to eliminate that gap and then it would swivel so you could lower it. The fix would be to put a small shim under the loop on the boom bracket so it wouldn't slide down when the boom is raised. Otherwise, I think my design will work really well.

I also got a picture of the mount I made for my anchor light today. Below is a picture of the bottom of the light showing the wire access to the light. The reason I decided to make a mounting bracket instead of just screwing the light to the top of the mast was so I could use the wiring access hole in the bottom of the light instead of cutting out the blank in the side of the light for wiring. There is also an issue with the ventilation of the light. There is a very small gap around the bottom of the light that provides for air circulation and if I

were to screw the light down on top of the mast, I'm afraid I would jeopardize the ventilation. With the metal mount, I can screw the light down without doing that. I used a piece of 0.090 inch thick aluminum plate to make the mount out of and then painted it white to match the mast. To the left is a view of the mount setting on the edge of the bench, showing how it will be attached to the mast. I will leave about a half inch gap between the top of the mast and the bottom of the light to allow room for wiring.

Saturday, August 03, 2013

I got my hinged mast step back from the welding shop and I spent a few hours getting it polished. It didn't polish up as well as the mast boom hinge did because the angle must have been extruded so there were a lot of really deep marks in it. I worked pretty hard on it and this is about as good as I could get it. (Picture to the right) The small box inside the two angles is attached to the bottom of the mast and the two angles will be attached to the deck. The bolt in the box passes through the angles and the mast, locking the mast in the upright position. That along with the forestays the mast will be supported in the

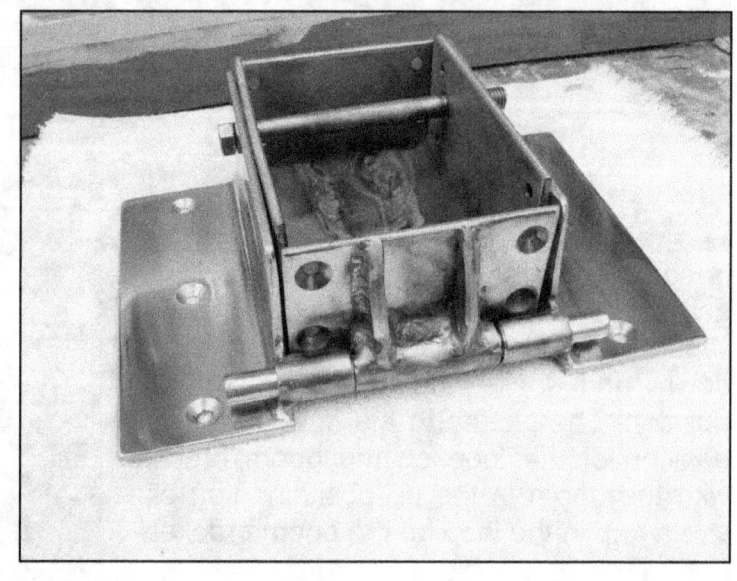

upright position. Removing the locking bolt and using the forestays, I will be able to lay the mast down if I run into a vertical clearance problem. Then I can use the forestays to raise the mast back up when I'm clear.

To the left is a view of the step box position with the mast laid down. This is about the position of the box but it won't be tilted as far as it is in this picture, it will be closer to straight up and down.

Sunday, August 04, 2013

Today I took some pictures of my mast hardware. Below is a picture of my boom swivel. It's a little hard to see, but the fix I mentioned before is complete. You can see a piece of filler rod in the loop that is

horizontal in this picture. This keeps the loop from slipping down when the boom is raised, locking the boom in the raised position. To the left is a picture of the swivel slipped over the end of the boom. You can maybe see the filler rod better in this view.

To the right is a picture of my mast step slipped over the end of the mast. In this view, you're looking at the front of the mast. This is where all the wires come out from the equipment mounted up on the mast. The only thing I will have to run outside the mast is the radar data cable because I have to keep that cable away from any other wiring. The fit was pretty good; I only had to do a little whittling to get it to slide onto the mast. The only problem I see is the positioning of the screw holes. I will have to use shorter screws for the front and back angles because the holes for the side angles are lined up pretty much with the holes in the front and back angles. I didn't see that coming when I drew up the design, but I don't think it's going to be too big of a problem.

Thursday, August 08, 2013

As I mentioned earlier, back in early April, I installed my air horns and they really sounded great. The only problem was that the air storage tank wouldn't hold pressure. The compressor would pump the pressure up to about 108 psi. Then after about 10 to15 minutes the compressor would come on again because the pressure had dropped to about 98 psi. I have spent the past 4 months, off and on, trying to figure out where the air leak was. I have replaced the air pressure gage, most all of the fittings on the air lines, most of the air lines and still the pressure bleeds off. I have called the people I bought the horns from a couple of times and they kept giving me suggestions as to what to try next, and nothing I tried stopped the leak. I finally decided I needed to isolate the solenoid valve which was mounted on the horns themselves to see if it was leaking. I thought if I took it off the horns and mounted it on the pressure tank, none of my air lines would have pressure on them until I pushed the horn button to blow the horns. I mentioned this to the people I bought the horns from and they said no you can't do that because the solenoid wouldn't work if I mounted it to the pressure tank. I couldn't understand their reasoning because if I hook the thing up the correct way, it should work the same as it does on the horns. So, I did that and I still had the leak. That told me that there was one of possibly three reasons for the leak. The solenoid was leaking, the pressure switch was leaking, or the tank itself was leaking. I purchased a new solenoid and a new pressure switch and installed the new solenoid this morning. I turned the compressor on and it pumped the pressure up to its normal 108 psi. I watched as the pressure began to bleed off so I crawled under the cabinets again and soaped every fitting I could find on the pressure tank. I found no leaks. I crawled back out from under the cabinets and looked at the pressure gage and it was holding about 2 psi below where the compressor shut off. When I left the boat, about three hours later, the pressure was still holding. If the pressure holds overnight I will be satisfied that I won't have to listen to that compressor every 15 minutes. I guess it was the solenoid after all but it took me a long time and a lot of frustration to figure it out.

Tuesday, August 13, 2013

I haven't done anything new lately so there isn't anything to take pictures of. I've been working on my mast and mast boom on the workbench and I have them almost ready for the final coat of paint. I put the second coat of primer on the boom today, and the mast is done and ready for the topcoat. I have finished pulling the last of the electrical wires through the mast and I only have the TV coax and the VHF antenna cable left to pull and I can close it up as soon as I get it painted. I need to get the bilges cleaned and ready for the last coat of paint in the engine room and the steering locker finished and ready for the last coat of paint and then I will mix up my paint and do all of them at the same time. Once I open the part B of my paint it is only good for a couple of weeks so I need to get everything ready before I do that.

I checked my clamps in the engine room and found quite a few that weren't all stainless steel even though they said they were all stainless when I bought them. Many only had a stainless steel band and the screw was mild steel so I decided I had better replace them with all stainless clamps. I found some of the clamps had rusted already around the screw and I found one that had actually come loose already. I spent most of today changing out the clamps that weren't all stainless. I still have about twelve left to do, but I couldn't find them locally so I had to order them on line. As soon as they arrive, I can finish the replacement of the rest of them. I also finished installing the aft braces for my thrust bearing mounting plates today as well. Now I have the mounting plates braced fore and aft so they aren't going to go anywhere no matter how much of a push or pull load is put on them. I have also finished assembling my fuel transfer panel. I cleaned the pump compartment so I could install the panel and not have to remove it again for cleaning behind it. I only have two headliner panels to install in the engine room and everything is done down there. Then I can lightly sand all the bilge areas and get it ready for the last coat of paint. It really doesn't need another coat, but there are several places where I have dropped a tool or something and dented the bilge planks a little which breaks the paint surface and that will allow moisture to penetrate into the wood so I decided I had better give the whole bilge area another coat to seal any breaks in the paint. I have also started putting diesel fuel in the boat. I was going to have a truck deliver the fuel all at once, but because I'm not using it for agricultural purposes I have to pay tax and delivery charges on it even though it is off road diesel. That adds up to more than I can get it for at the local gas station. So, I decided to just get my gas cans filled whenever I'm in town and fill it that way for a little less money.

I'm really hoping it's going to cool off down here in Florida pretty soon because once I get the everything painted, I need to start working on the fly bridge to finish it up and right now it is well over 125 degrees by noon up there next to the roof of the building. You just can't work up there right now. I have had to go up there once or twice to pull a wire or something, and it is so hot it actually takes your breath away, it's overwhelming.

Saturday, August 24, 2013

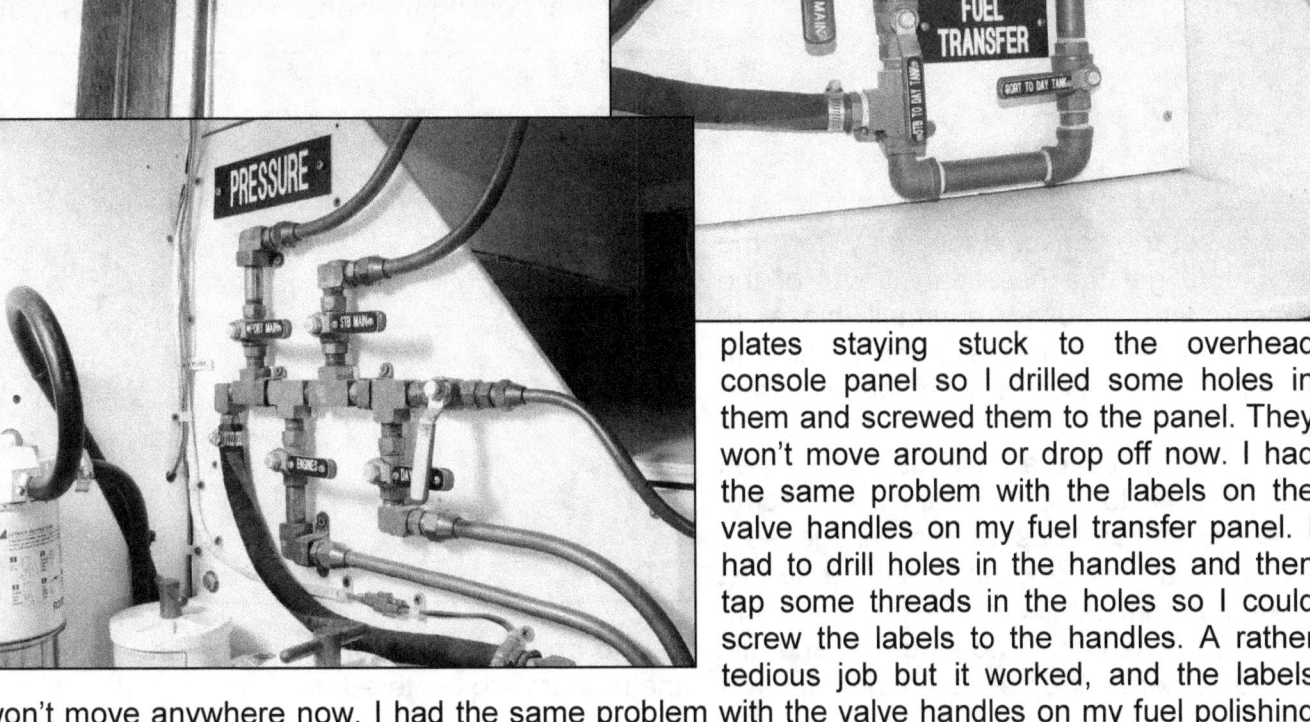

I finally got some pictures of what I've been doing lately, nothing much new, but just cleaning up loose ends. It's a little hard to see in this picture below left, but I had some trouble with these little name plates staying stuck to the overhead console panel so I drilled some holes in them and screwed them to the panel. They won't move around or drop off now. I had the same problem with the labels on the valve handles on my fuel transfer panel. I had to drill holes in the handles and then tap some threads in the holes so I could screw the labels to the handles. A rather tedious job but it worked, and the labels won't move anywhere now. I had the same problem with the valve handles on my fuel polishing

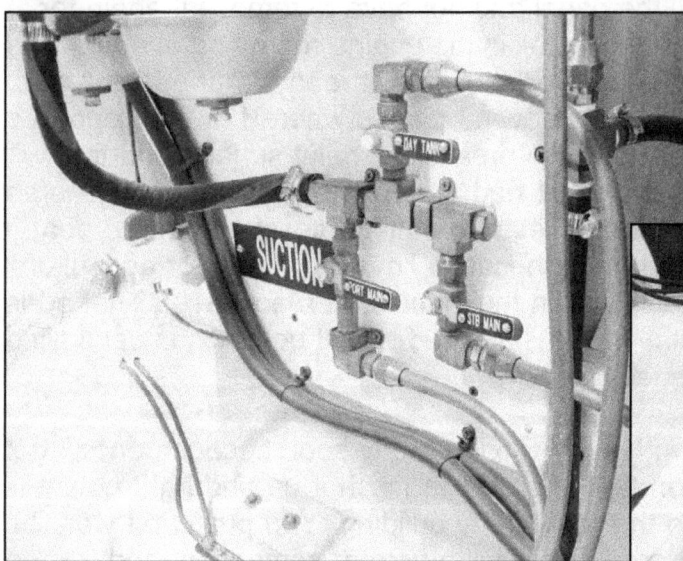

panel; the picture to the left is the pressure side of the panel. I screwed the labels on them as well. To the left is a picture of the suction side of the panel. Sure are a lot of valve handles to drill and tap for those labels.

The picture to the right shows the shaft seal cooling hose which is attached to the engine raw water system and the crossover hose that goes to the other shaft seal. This will allow one engine to cool the seal when the other one is shut down.

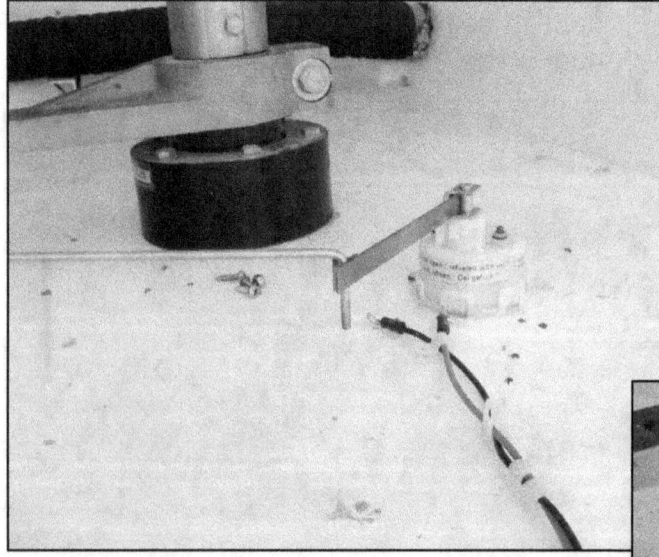

The picture to the left shows my sensor for the rudder angle indicator gage. I have spent endless hours trying to figure out how to hook this thing up. I couldn't seem to get enough travel out of the sensor to match the travel of the tiller arm. Then I tried attaching the sensor via some linkage to the tie bar, and I finally got it to work. The picture to the right shows the linkage connecting the sensor to the tie bar. I really

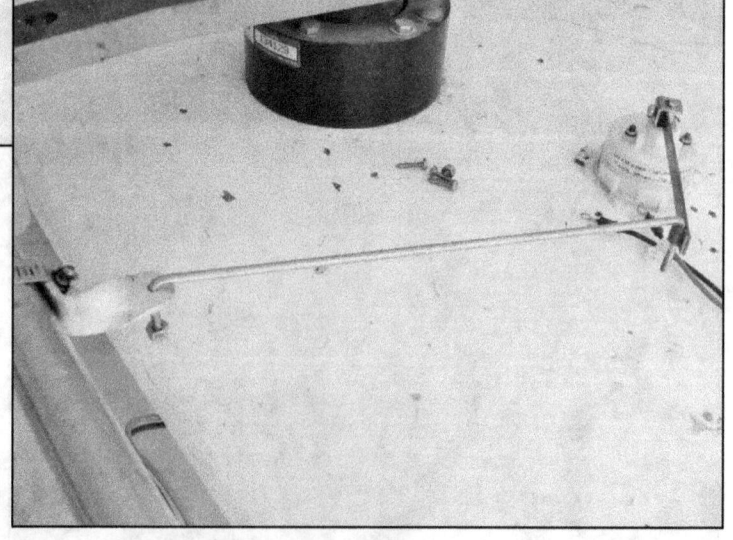

had to set the sensor a long way from the tie bar to get the necessary travel of the sensor arm. I used a small piece of aluminum flat bar to make the bracket to attach it with. I bent it to 90 degrees and twisted it so I could drill a hole in it for the linkage. I just used a hose clamp to fasten the bracket to the tie bar. I cut some threads on the ends of the linkage so I could put a fiber lock nut on each end just to keep it from coming off the bracket or the arm on the sensor. It works great, and all I need to do yet is hook up the wires and adjust the sensor to center the needle on the gage when the rudders are centered.

To the left is a little closer view of the little mounting bracket I made. I reduced the width of the flat bar where it fits against the tie bar so the clamp would hold it straight against the tie bar.

I've started building the shelves for the steering locker. To the right is a picture the first one laid out on the work bench. This one will go right on top of the rudder shelf. As you can see, I put a 1x4 all around the outside of the shelf. This will prevent of anything from sliding off the shelf in rough weather. The picture to the left is another view of my shelf which shows the notch I had to make so the shelf will clear the rudder angle sensor. It was fun making the edging of the shelf go into the notch. Now I have to take it all apart and carry it up into the steering locker to see if it will all fit before I glue and screw it together. Then it will be one shelf down, seven to go. This may take a while.

Saturday, August 31, 2013

I'm still working in my steering locker building some more shelves. To the right is a picture looking down into the steering locker through the hatch. You can see the two shelves on the forward bulkhead. These are a little smaller in width so as not to interfere with getting in and out of the locker. The rest of the shelves are 1 x 12's so they will accommodate more than these little 1 x 8 shelves. I have all the shelves built and installed now. All I have left to do is support them with some braces because some of them are pretty long without any support under them. When I get the bracing done, I will remove them and put them on the workbench so I can put a coat of primer on

them and then paint them before I reinstall them. It will be a lot easier to paint the steering locker without the shelves in there. To the left is another view of the two 8 inch shelves. As you can see, I put a 45 on the corners to eliminate sharp corners. Below is another

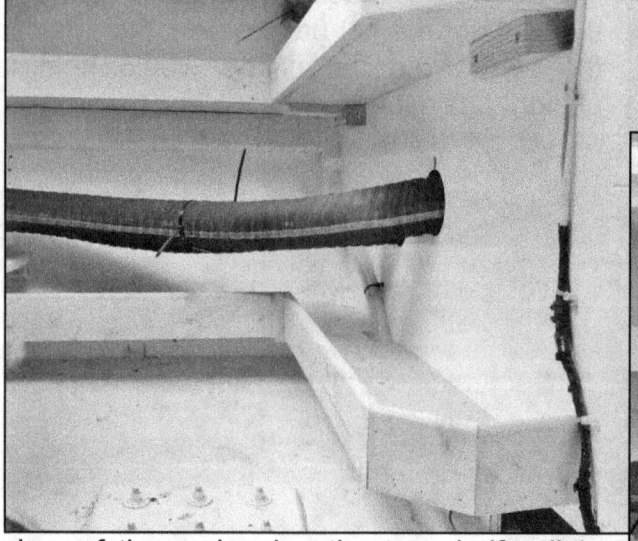

view of them showing the top shelf a little better. Below is a view of the corner of the

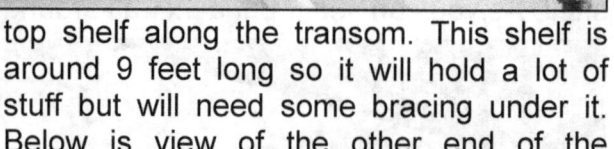

top shelf along the transom. This shelf is around 9 feet long so it will hold a lot of stuff but will need some bracing under it. Below is view of the other end of the

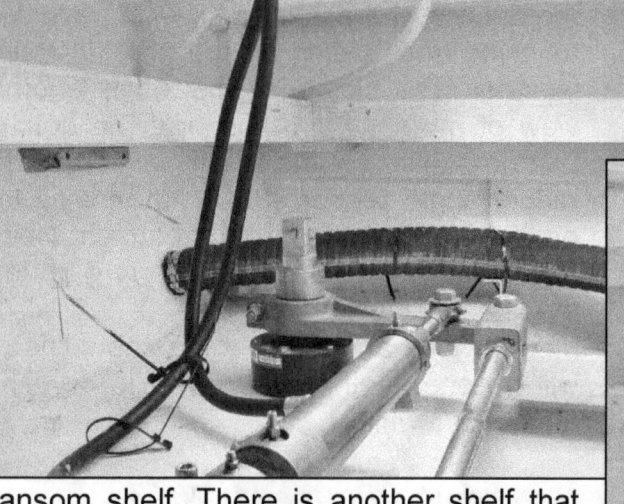

transom shelf. There is another shelf that goes on top of the rudder shelf which is not

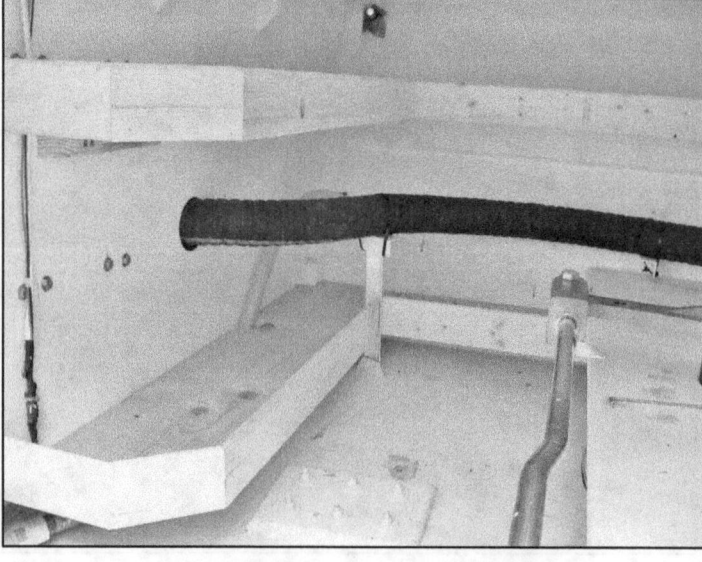

in this picture because I have it out for painting. That is the one that is on the workbench on the previous page. To the left is a view of the two shelves along the forward bulkhead on the starboard side of the steering locker. I should have plenty of storage space with all these shelves for my spare parts, tools, extra oil and so forth. These shelves take up a lot of valuable space but I need them to maximize the storage capacity of the steering locker.

Wednesday, September 04, 2013

I have all my shelves removed and set up on the workbench. I have primer on all of them now and they are ready for the polyurethane. Below are a couple pictures of the shelves on the workbench. This is almost as bad as when I had 25 raised panel cabinet doors on the workbench. As you can see, there are quite

a few pieces to these shelves. I have them all labeled so I hope I can get them reinstalled in the right place.

Monday, September 30, 2013

It's been quite a while since I have updated my journal because I've been sanding and cleaning getting everything ready for paint. I pretty well have all that done now, along with a few other little projects along the way so I thought I had better get busy and update.

One little project along the way I needed to get done was to install my isolation transformer because I have requested a marine electrician to come and wire it for me. I have read everything I can find on the subject and still can't figure out how to do it. I'm having trouble understanding the bonding of the AC system and how that affects the wiring of the transformer. I think I have it figured out but I'm just not sure so I thought I should call the experts. To the left is a picture of the transform installed in the

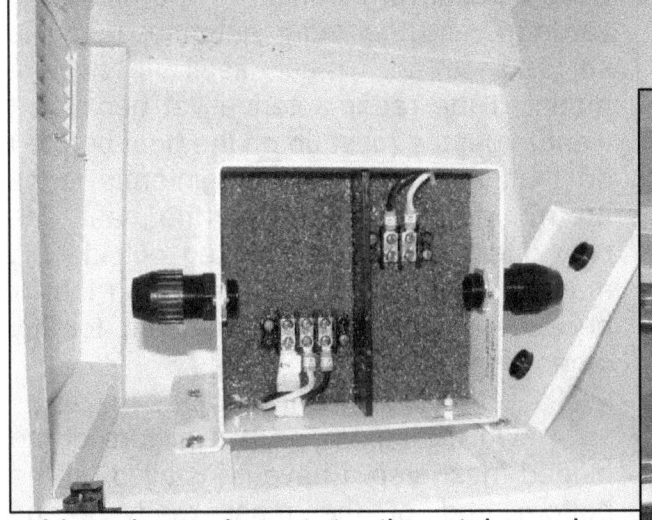
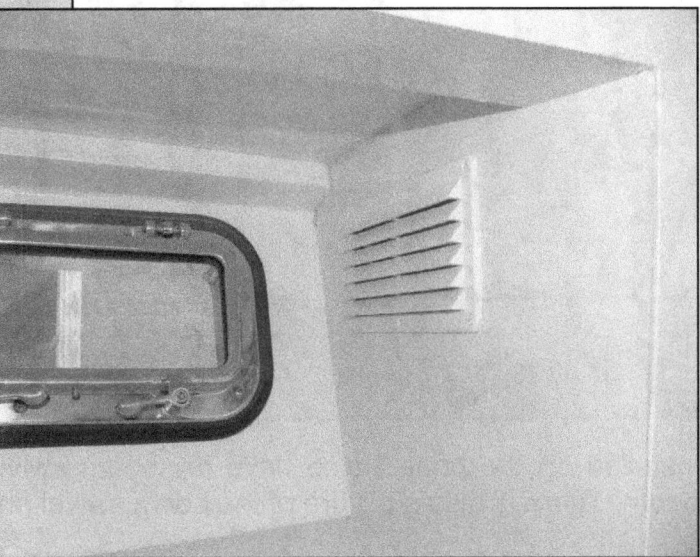

cabinet located next to the stairs going down into the forward cabin. I chose this location because it is right under where the shore power comes into the boat. The

transformer does produce some heat, so I needed to ventilate the compartment. I installed a couple of vents, one in the door, and one in the side of the compartment to allow for circulation of the air through the compartment. The vent on the side of the compartment is pictured on the previous page and the vent in the compartment door is pictured to the left. The location of these two vents should provide for natural air circulation through the compartment to keep the transformer a little cooler. Now that I have the transformer installed, I guess I will have to get started on my painting. I can't think of anything else I need to do right now so I guess I will have to start painting, one of my favorite jobs!

Tuesday, October 15, 2013

I have finally finished putting my mast together. To the left is a picture of all the equipment mounted on the mast and the yardarms. I had a lot of trouble mounting the TV antenna on the port yardarm. The rail mount just wouldn't hold the antenna very well so I had to make a small extension out of brass to secure the antenna to the rail mount. It is very secure now so I don't have to worry about that. I also have the VHF antenna mounted on the back of the mast, and the navigation and anchor lights mounted on the mast. The GPS antenna is mounted to the starboard yardarm, and the radar antenna mount is secured to the mast itself. I haven't mounted the radar antenna yet because I want to get the mast up on the boat before I do that. All the wiring and antenna leads are run through the inside of the mast. The only one that will be mounted outside the mast is the radar antenna lead and that is because it can't be run adjacent to any other wiring.

On the next page is a picture of the hinged mast step. I have it installed on the bottom end of the mast ready to be installed on the boat. I also have my boom swivel mounted to the mast, ready to accept the boom. There is also a picture of the boom swivel on the next page.

To the left is the picture of the hinged mast step. You can see all the wires for all the lights and antennas coming out of the bottom of the mast. Below is the picture of the boom swivel mounted on the mast.

Saturday, October 26, 2013

I've been working down in the steering locker the past few days, and I finally have it all completed. I'm sure glad to be done down there, because there just isn't enough room in that place to turn around, literally! I have all my shelves installed and braced and I also installed a couple lights down there because it is a little dark over on the starboard side of the locker. Once I finished in the steering locker, I started on the wiring of my isolation transformer. That is the thing that had me really stumped but I contacted the supplier and he explained a few things to me so now I think I can get it wired correctly without hiring an electrician. It really turned out to be simple enough, I just couldn't figure out which way to wire the thing since there is two different ways to do it. I should have that done in a couple more days, and then it's on to more loose ends

Sunday, November 03, 2013

I still don't have my isolation transformer installed. I ran out of wire, so I had to order another two feet of the 6/3 before I can finish hooking it all up. It shouldn't take too long once I get my wire. In the meantime I have been working on a few loose ends on the bridge and in the salon. I had to put some cabinet door latches on the doors under the bench seat on the bridge and I also close everything up in there so I think I am done with that part of the boat except for a final cleaning. Then I finished sealing the decks in the forward cabin area and I had one more coat to put on the salon deck so I got that all done. Then I moved to my entertainment center where I needed to put a couple latches on the doors there. I ran out of latches (that's par for the course) so I decided to install the antenna booster. That turned out to be a bit challenging because the coax inside the cabinet wasn't long enough to put the ends on outside the cabinet. I finally got them all installed and then I had to run some wire for the DC circuit that powers the booster. I don't have any DC in the salon except for the lights in the overhead so I had to run some wires all the way from the forward cabin to the entertainment center, and that took a while. I should be able to finish that up tomorrow and then I think I will work on my anchor rode. I'm going to mark it so I can tell how much I let out when anchoring. I just have to pull it all out of the anchor locker, measure it, and mark it. That shouldn't take too long so I will have to find something to do when I get that done. I'm getting close to running out of things to do.

Saturday, November 09, 2013

My wire arrived and I finished installing my isolation transformer. It came on and started humming when I through the switch so I guess I got it installed correctly. Everything is still working as it did before so the AC is getting to the main distribution panel like its supposed to. Then, I pulled all the anchor line out and marked it every 20 feet because I have 20 feet of chain so I just continued marking at 20 foot intervals. Now I just have to watch the markings as I let out the anchor and I will be able to tell how much rode I have out so I get the right scope for the depth of the water.

I filled the fresh water tanks with water, and turned on the water pump. I started checking for leaks, and it seemed like the pump wouldn't pick up the water from the tank. The pump kept pulling air from somewhere but I couldn't figure out where. When I turned the pump off, I saw a dripping of water at the output side of the filter. That's where the pump was picking up the air. I couldn't see the leak as long as the pump was running. I also had a slight leak at the galley sink drain which I fixed with a little tightening of the drain connection to the trap. Then I checked all the rest of the sinks and the shower and everything seemed to be good. Then I noticed quite a bit of water in the bilge. I pulled the hatch cover over the sump pump and it was really wet. My sump pump wasn't coming on and the tank was full and overflowing. Once I turned the sump pump breaker on the water in the tank was pumped out and back to the gray water tank. I was sitting in the engine room and I could hear the pump coming on just slightly every so often. So, I figured I must have a leak somewhere; I just had to find it. I finally found the leak in a tee fitting under the galley sink in the hot water line. I got that fixed and then I noticed a leak dripping off the cold water line where it runs through the deck to the galley sink. I checked the lines under the sink again and they were all dry, no water anywhere and that is the line that the water is dripping off below the deck. I haven't found that leak yet, but I will have to keep looking until I do. Other than that, everything in the plumbing system seems to be functioning as advertised. I just have to find the last leak.

Tuesday, November 12, 2013

I've spent the past three days chasing water leaks and I finally have them all fixed. I think my hoses must have dried out and shrunk because almost every hose clamp was loose but once I tightened them, the leaks stopped. Now I have pressure on the whole system and it holds the pressure overnight without any leaks anywhere.

Once I had my leaks fixed, I tried to get my water heater to work. Needless to say, I can't seem to get it to fire off. I know I have gas in the line, because I have lit all the burners on the cook top and they all work fine. I couldn't seem to get enough water flow through the water heater because you need at least 0.4 GPM for the heater to come on. I removed the water pump and took it apart to see if there was anything clogging the pump and it was fine. Then I took the restrictors out of the faucet in the head and that gave me a little better water flow. Now I can get the heater to come on when I turn the hot water on, the fan anyway, but it still won't light off. I'm still not getting enough water flow with just the galley sink faucet turned on, so there must be a restrictor in that one somewhere but I can't find it. I can't figure out how to take the aerator off which is where the restrictor usually is. I took the front panel off the water heater and I have a flashing green LED which means the system is operating normally. Next I think I will have to crawl under the sink again and see if I can look into the igniter hole to see of the sparker is working. I hear a click when the fan starts up which I think is the sparker but I don't know if I'm getting a spark or not. That is the only thing I can think of at this point that would keep the heater from firing off. I'll have to check that out tomorrow. Then I need to figure out how to remove the restrictor in the galley sink faucet and I should be good to go, hopefully that will give me enough flow to start the water heater, if not, I don't know where to look next other than a restriction in the hot water line somewhere. Good luck finding that! Building a boat sure is a lot of fun!!!

I know all this text is a little boring but there just isn't anything to take pictures of while I'm working on testing these different systems. I will post some more pictures as soon as I can.

Friday, November 22, 2013

I finally have something to take a picture of. I have been working on my winch mount for the past couple days. Here are a couple pictures of my progress to date. To the left is a picture of the bottom frame partially assembled and welded. To the right is a close up view of the threaded hole in the left side vertical member that I will use to attach the mount to the mounting brackets which will be secured to the deck.

November 26, 2013

To the left is a view of the frame with one of the winch mount plates positioned so I can tell where to weld the cross members. I'm having fun welding the stainless steel angle with my oxy-acetylene torch. The experts at the local welding supply said I couldn't weld stainless with my torch, and I would have to us TIG which I don't have. A friend of mine who owns a metal shop said he didn't know why you couldn't although he had never tried to do it himself, so he call another friend who was a welder and he said he does it all the time. Then I found a video on line that showed how to do it so I am doing it the same way and it works very well. There is a trick to it, but once you master the process, it's almost like welding mild steel. You just have to use a carbonizing flame so you shield the molten metal and the stainless steel TIG rod (308) you're using to fill with and the acetylene rich flame keeps the metal from oxidizing.

I'm making some pretty good progress on my winch mount. Here is another picture of where I'm at right now. I have the bottom frame all welded into the vertical pieces. You can see the winch mount plate and fair leads that bolt to the winch and is in turn bolted to the mount I'm making. I ran out of oxygen this afternoon so I had to quit and go get another bottle. Hopefully I can get it finished up tomorrow.

December 26, 2013

I hope everyone had a very happy holiday, and yes I'm still working on the Molly B even though it's been awhile since I have posted anything on my site. My winch mount is coming along very well and I have it completed now. All I need to do now is try to get it up onto the boat deck so I can mount it. This thing weighs close to 100 pounds with the stainless steel frame and the two winches mounted in there.

These two pictures show the winches mounted and the one to the right below shows about where it will mount relative to the mast.

Here are a couple pictures of the finished product. To the left is a view of the front of the mount with the box I put around it to keep the winches out of the weather.

To the right is a view with the front panel installed and shows where the two cables come out through the front panel. You can see a large handle on the top of the box in the picture to the right. I hope that will facilitate getting it up onto the top of the boat.

January 2, 2014

Happy New Year everyone!! Another year has passed and still the Molly B is under construction. I am getting close to the launch now, and it shouldn't be too much longer. My oldest son came down from Orlando the other day and helped me get my mast and winches up on the top deck. Here are a couple pictures of them. To the left is a view of the mast and winches with the back part of the fly bridge on its side so I could do the wiring easier. To the right is a view of the mast with the back portion of the bridge upright.

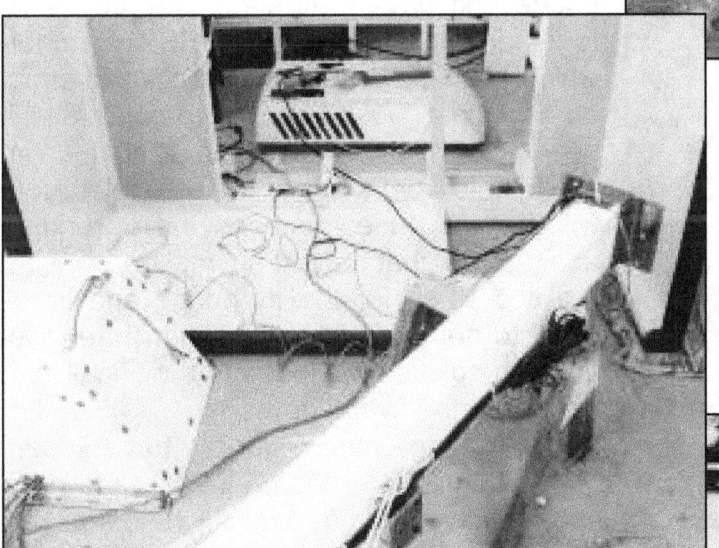

Here are a couple views of some of the wiring. Below is a view of some of the wiring under the bridge dash. Some of this wiring will remain with the bridge when it is removed, and some will remain with the boat. It just depends how they are routed up to the fly bridge from below or from the main bridge. It's a little confusing, but I have them separated and labeled so I know which ones are which.

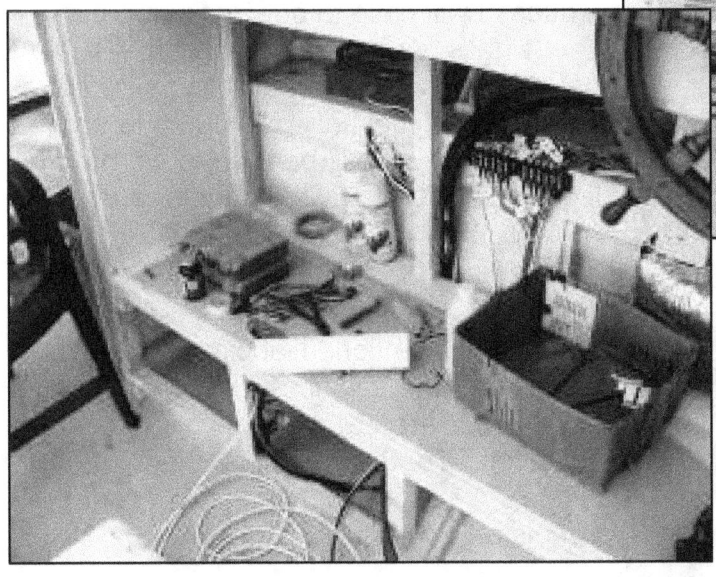

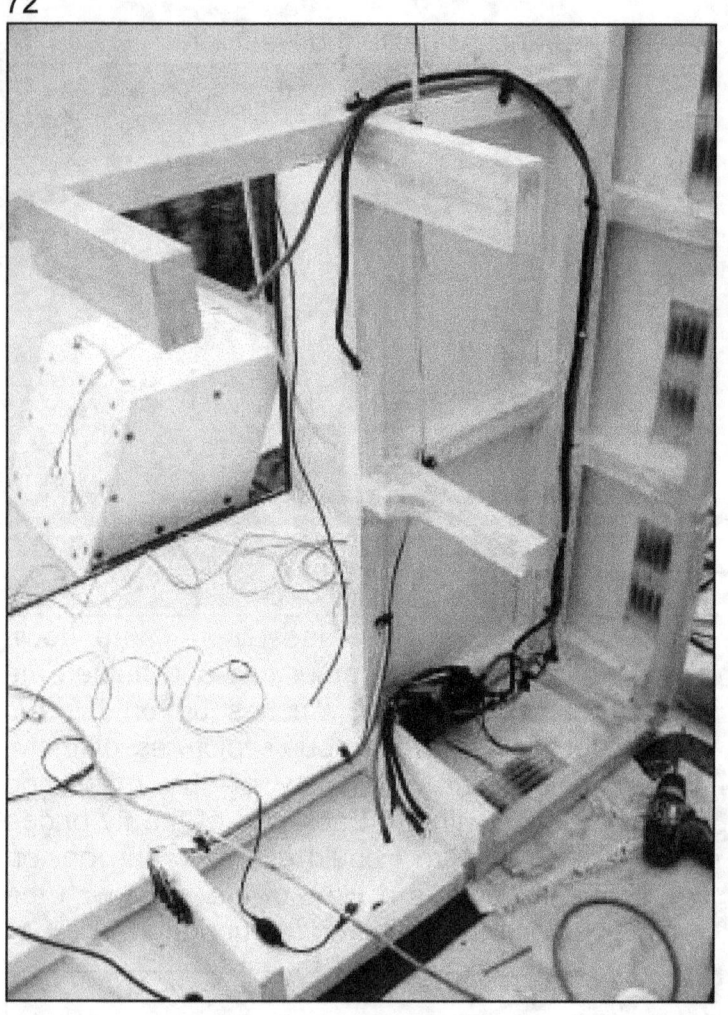

To the left is a view of the wiring attached to the underside of the aft portion of the bridge. This wiring will remain with that portion of the bridge when it is removed for transporting the boat to the water. This is the wiring for the two winches. The two little boxes at the bottom of the picture are the solenoids for the winches. The switches on the mast will operate these, and in turn they will operate the winches.

Thursday, January 30, 2014

I've finished the wiring on the fly bridge and have it all pretty well put back together. I have ordered my AIS system for which I will need to install another antenna on the mast. Just one more coax cable I need to run to the black box and then a USB cable from there to the computer, so I will probably need to open the fly bridge again to get that done. I have also received the two seats for the fly bridge and I will need to get them installed. Then I need to finish the glove box door and I will finally be done with the fly bridge.

In the meantime, I have been working on building the valances for the windows in the salon. I have them all built, and now Linda will attempt to cover them with upholstery material. Then we need to decide whether to put curtains or blinds or both on the windows.

I have also been checking my fuel systems in the engine room. I had a couple leaks that I have fixed and I have one that is just seeping a little but to fix it, I need to transfer the fuel out of the port main tank and into the day tank. I attempted to do so with my fuel transfer pump and the pump didn't seem to be able to pick up the fuel from the tank. While trying to get it to pump out the tank, it seized up and stopped working altogether. I guess I will have to pull that pump out and see what's wrong with it. I hope I won't have to replace it like I did the water pump. I can use my fuel polishing pump to transfer the fuel so I will still be able to work on the leaks while I try to fix the transfer pump.

Sunday, February 2, 2014

I guess I had better get this website up to date. As I mentioned above, I have been working on the valances for the windows in the salon. Here are a couple pictures of the valances. To the left is a view of them sitting on the bench, all ready for upholstery.

To the right is a view of the trim piece that will go on each valance over the fabric covering. There will be a rope attached to these trim pieces to make them stand out from the fabric covering.

I found out what was wrong with my fuel transfer pump once I got it torn out and on the work bench. I checked the pump end and it was fine, and then I noticed a little sticker on the opposite end of the motor that said "Fuse Inside". What a dumb place to put a fuse. I removed the end cap from the motor and sure enough, the fuse was blown. I replaced the fuse and the pump ran fine. I rewired the fuse to put it outside the motor end cap and installed the pump. I checked the operation and it pumped fuel from the port tank to the starboard tank and back to the port tank again. I then pumped the fuel from the port tank into the day tank and everything worked great. Then I transferred the fuel into the day tank. It was then that I discovered my day tank fuel gage didn't work. I checked the terminals and I was getting voltage from the sender to the gage but it wasn't reading the contents of the tank, so I decided it must be the gage that wasn't working. I have a new gage on order so as soon as it gets here I can get it installed.

Wednesday, February, 26 2014
It's been few weeks of nothing but setbacks!

I finally found some ISO 10 hydraulic oil for my steering system after about two weeks of searching, so I decided to fill and bleed that so my rudders would be working. I bought two gallons of oil and used all that the first attempt and I still needed more to fill the upper helm pump so I ordered another two gallons. I got the third gallon poured into the pump and I still couldn't keep the pump full so I decided there must be a leak somewhere. I went down into the raised pilot house and there was at least a gallon of oil all over the aft bulkhead the seat, and the two compartments below the seat. What a mess! I checked the only connection in the pipe running from the upper helm pump to the lower helm pump and it was a little wet, but didn't seem to be leaking badly enough to generate a spill of this magnitude. The only other possibility was that there must be a hole or crack in the pipe itself which was inside the bridge roof. I tour the aft headliner down and found a screw that had been screwed right through the top of the pipe and out the bottom. That was my leak and I had to cut out the pipe with the hole and patch in a new section of pipe to fix the leak. I got it fixed and the bridge all put back together finally and my rudders are working great now.

I had my rudder angle indicator sender all hooked up and working but I realized that the gage was reading 180 degrees out. I tried everything to get it to read correctly and the only fix was to actually turn the sender around 180 degrees and then it worked correctly. The only problem then was the fact that the linkage connecting the sender to the tie bar, didn't work right as it would go fast in one direction and slowly in the other direction, so the geometry was off. I had to make a new linkage to make it work right. I finally got it fixed today.

Since I had fuel in the boat and I finished checking the fuel systems and pumps, I decided to hook up a water hose to the engines and run them to check everything out. I hooked up the hose to the port engine and the fitting on the raw water pump just sprayed water everywhere. So I decided to check the starboard engine and it seemed to be fine so I went up to the bridge the try to start the starboard engine and the battery was dead so I had to put the charger on it. In the meantime I decided to remove the raw water pump from the port engine and see if I could fix the leak. When I got the pump off I noticed some small pieces of black rubber in the fittings. I opened the pump cover plate and the impeller was totally deteriorated. Since the port pump impeller was shot, I thought I had better check the starboard pump and it was bad as well. I got new impellors and installed them but then I couldn't find the right gasket for the pumps. My local Perkins dealer found them for me in Jacksonville and now I have them ready to install except that I can't find a hose barb with a straight thread to fix the leaking problem. I'm still working on that.

Sunday, March 02, 2014

I finally finished repairing and installing my two raw water pumps. I decided to run the engines so I hooked up the water hose to the starboard engine and hit the starter. After a little cranking, it started and ran great. I had good oil pressure, water temp was good, and the alternator was putting out good voltage. I ran it for about a half hour and the water temp held right around 180 so everything was good. I checked the engine and found no fuel, water, or oil leaks. Then I hooked the water hose up to the port engine, and when I hit the starter switch, nothing happened, not even a click from the starter solenoid. I knew the battery was good, because I had just charged both starting batteries. I took the instrument panel out and found a disconnected wire on the start switch. I hooked it back up and hit the starter. The engine cranked but wouldn't start. I went down and checked the wiring on the injector pump solenoid and found a wire disconnected there. I hooked it back up and when I hit the starter the engine started and ran great. After about ten minutes the engine died and wouldn't start again. It acted like it just ran out of fuel. I am now in the process of tracking down the problem but haven't found it yet.

I've also been working on the handrails on the boat deck. To the left is a picture of the hand rails along the port side. They are a little hard to see from this distance, but as you can see, I ran out of stainless steel tubing about half way up the side. I have the remaining tubing on order, and I should be able to finish it up when it gets here. Below is a view of the back of the deck where I have the rail completed. I haven't screwed the stanchions down yet because I will have to remove all the rails when I move the boat to the water. I'm concerned that if I screw them down I will have holes in the deck that would allow water to enter the plywood up there if it rained after I get the building torn down and during transport of the boat to the water. I might be able to seal the holes with some caulking to prevent this but I'm just not sure what I want to do yet. I would like to be able to secure the handrails just to make sure I have them all cut and fit properly before I move the boat.

Thursday, March 13, 2014

After checking everything on the port engines' fuel system, I came to the conclusion that it must be the lift pump. I was going to replace it with one of my spares when I decided to remove the fuel solenoid from the starboard engine and put it on the port engine just to make sure that wasn't the problem even though I could hear the port solenoid click when it was activated by the switch. With the solenoid from the starboard engine installed on the port engine, it only took a little cranking and the engine started and ran great. Problem solved! I ordered two new solenoids so I would have a spare. I had a friend bring his son over to look at the boat and it turned out that he was a diesel mechanic. After explaining all the trouble I had with getting the engines running, he wanted to hear them run so I got a "Y" valve and another hose and hooked both engines up to the water supply. We cranked each engine and they both started right up. He said they really sounded good. While he was there, he went down and set the idols so the engines were synchronized at idol and then again at full throttle.

I've also been working on my blinds and valances in the salon. It took a few days but I finally finished installing them all today. Below are a few different views of the windows in the salon with the blinds and valances installed.

As you can see, I have them all installed. The blinds were the hard part although covering the valances wasn't easy. I had to shorten every one and that was a real pain. I also had to cut the top and bottom rails down to allow for the radius at each corner of the windows.

Saturday, March 22, 2014

Since I've completed the window trim in the salon, I have started working on the ladder rails for the boat deck. I have the railing completed for the opening in the deck for the aft ladder, and I have the hand rail for that ladder completed as well. I have a good start on the railing for the forward ladder and should be able to finish it up in the next day or so.

To the left is a view of the handrail I'm building for the forward ladder to the boat deck. Below is another view from the

side that shows the railing a little better. I have it all finished now and ready to be painted and installed.

Tuesday, March 25, 2014

I have my railing for the forward ladder all painted and ready to install. The railing for the aft ladder is also ready to install. I should be able to get them installed tomorrow and I will get a couple of pictures of them. I have also been working on the antenna for the AIS system and I have it all installed and the cable run down into the bridge. The AIS is done except for running the USB cable to the navigation station where my computer will be installed. It's supposed to be cool tomorrow so I think I will get to work on the handrails around the boat deck and get the stanchions screwed down. All I will have left to do up there then is finish the antenna and GPS cable attachments. Then I should be ready to wash down the deck up there to get the top of the boat ready to put the last coat of clear on the topsides of the hull. Linda has been cleaning the inside of the boat getting it ready to start loading our stuff in there. So, it won't be long and the building will have to come down in preparation for moving the boat to the water. It's been a very long 11+ years but we're almost done!

Wednesday, March 26, 2014

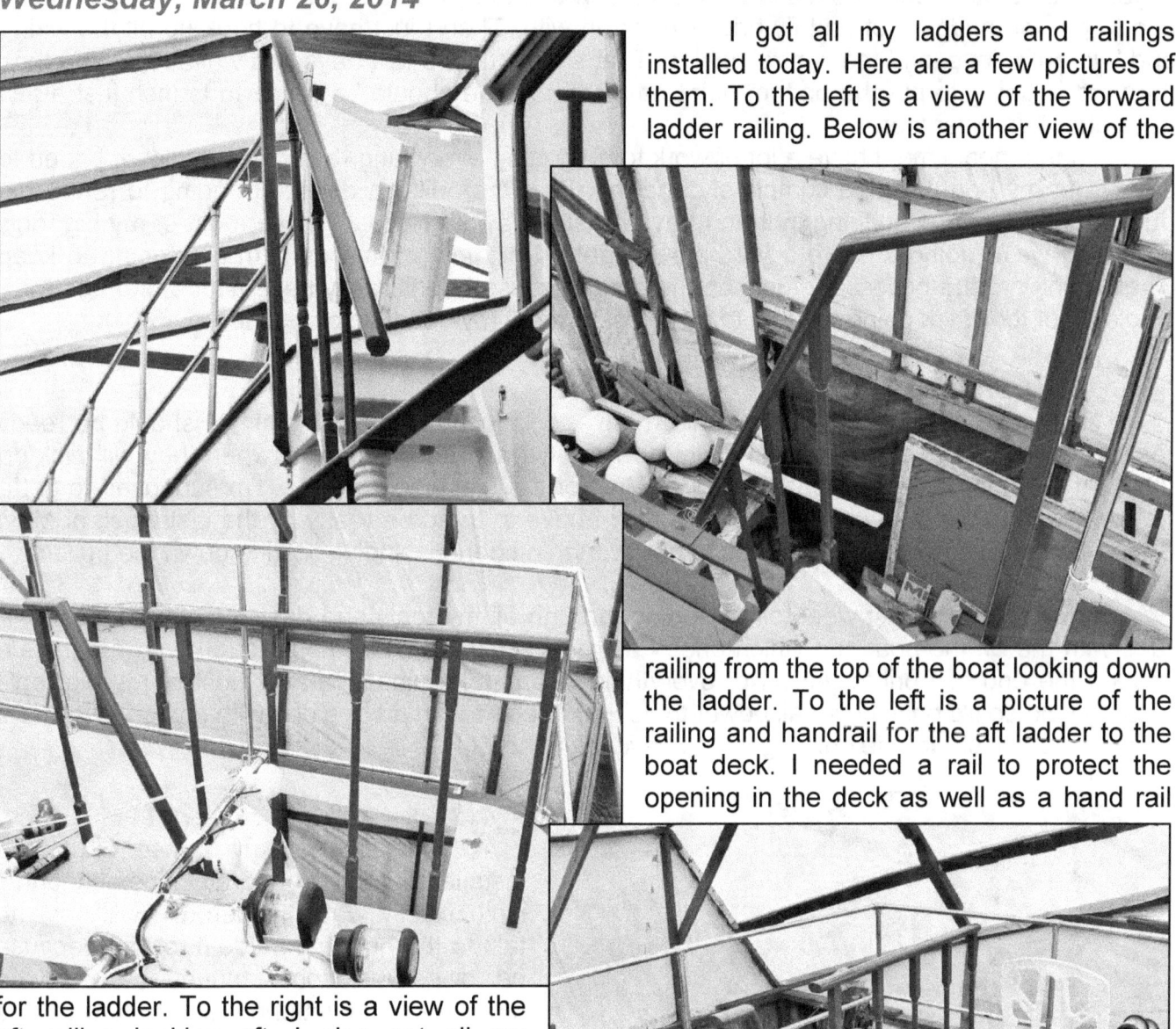

I got all my ladders and railings installed today. Here are a few pictures of them. To the left is a view of the forward ladder railing. Below is another view of the railing from the top of the boat looking down the ladder. To the left is a picture of the railing and handrail for the aft ladder to the boat deck. I needed a rail to protect the opening in the deck as well as a hand rail for the ladder. To the right is a view of the aft railing looking aft. I also got all my stainless steel handrails screwed down today. Once the stanchions are screwed down the rails are nice and solid.

Friday, April 04, 2014

From the pictures above it doesn't look like I'm anywhere near ready to move the boat, but actually I am. Most all of the building is complete and I have just finished putting the last coat of clear topside paint on the hull so all I have left to do is tear down the building and clean the rest of the boat up to get it ready to move. I have a deadline now because I have made arrangements to have the boat moved on the 1st of June. I'm moving it to Cracker Boy Boat Works in Ft. Pierce, FL. I really had a hard time finding a transport company to move the boat because most of them said their insurance wouldn't allow them to move a wooden boat. Don't ask me why, but that was the most common excuse I found. One said they wouldn't pick up a boat in a residential area, and some said they couldn't move a trawler because they were too high to transport. Anyway, I finally found one that will move it for me and a crane company that will lift the boat and load it onto the mover's trailer. The boat yard on the other end has a travel lift that I can use to unload the boat and a small crane I can use to put the bridge roof and fly bridge back on with. Then I just have to hook up all the wiring and hydraulic lines, and the control cables. That should take a few days and while I'm doing that, the yard is going to put the bottom paint on for me so we should be ready to launch just a few days after I get over there.

In the meantime, I have a lot of work to do getting everything ready for the move. I need to empty out my boat building completely and then tear the building down. I'm going to rent a big dumpster and load everything in it and have it hauled away. I already have most of my big tools sold, and we're going to have a yard sale to get rid of most of the rest. I'm only going to keep what I think are the necessary tools on the boat because I don't really have a lot of storage for a whole lot of tools. I'm going to have to work fast to meet my deadline of 1 June.

Monday, April 14, 2014

The days are going by too fast and I have too much to do but I think we should be ready for the move whenever it happens. Right now I only have an estimate of when they will pick up the boat and that is either the last week of May, or the first week of June. I'm supposed to get a narrower window in a week or two but for now I have to try to be ready by the last week of May. That only leaves me about five weeks to get everything done including tearing down the building.

I have spent the last week or so sorting through all my tools and deciding which ones I will take with me on the boat and which ones I am going to try to sell. I hope to have a garage sale around the end of April to get rid of everything I can't take along. I have most of my big tools already sold so it's only the small power tools that I need to get rid of along with my table saw and compressor which are the only big tools I have left to sell.

Wednesday, April 17, 2014

I finally got my transom lettering installed. It turned out pretty nice although it isn't exactly what I ordered. I will have to talk to the guy that made it for me because he owes me some money because he didn't make the lettering like I ordered it and paid for.

Sunday, April 20, 2014

The tool sale really went well. We sold all the tools and a lot of other stuff as well. I only had to throw away a few fittings and staples, the rest got sold. Now I can start working on taking the building down. That ought to be a lot of fun! Right now I'm working on tearing apart all my work benches and getting everything cleaned up inside of the building. I have a 30 cubic yard dumpster parked along the north side of the building that I'm going to throw everything into and have it hauled away. To the right is a picture that shows the plastic

being removed from the building. It looks like bad storms blew through and tour the plastic off the building. Actually my helper John and I just cut it off between the rafters and wall studs. To the left is another view of the building with the plastic removed. There is just a little more on the front part of the roof left to remove.

Wednesday, April 30, 2014

Below is a view of Norm, my neighbor pulling the end rafter off the roof. It got hung up on the ends by the plastic and wouldn't come down until we cut away the rest of the plastic.

To the left is a view of John and I removing the rafters from the back part of the building. This is not such an easy task since I can't let them just drop onto the boat; we have to take them down one half of a rafter at a time so we can handle them to keep them off the boat.

Thursday, May 1, 2014

"And the walls came tumbling down!" Once we got all the rafters off, we just pulled the walls down with a few ropes, one at a time. Some of them we sawed up with a chain saw, and some we just knocked apart with a sledge hammer and threw them into the dumpster. It didn't take long once we got to this point to get the building all torn down. I got the whole thing into the dumpster except for the south wall. I had to have them empty the dumpster and bring it back for the rest of the building.

This is the first time since I started building this boat that I have been able to get the whole thing in one picture. It gives you a better perspective of just what it looks like. I think it turned out pretty well. You never know if what you are doing will look good with respect to the rest of the boat if you can't stand back and see the whole thing. The view from the front really looks good. You can see the remnants of the south wall a little better in this picture. It's going to take a few more days to get all the rest of this lumber cleaned up and loaded into the dumpster. Then I can start getting the boat ready to be moved.

Thursday May 15, 2014

I have gotten quite a bit done in the past two weeks. I finished taking down the building and got it all hauled away. The rain seemed to continue just about every day since I got the plastic off the roof and I had several water leaks that made their way all the way down into the bilges. I couldn't figure out why I had leaks when it dawned on me that I only had the bridge roof and fly bridge sitting on the boat, and not sealed or screwed down and that is where all the water was coming in, around the fly bridge mostly. I had a gray tarp so I stretched over the bridge and fly bridge and I hope I have stopped the leaks. It's been raining almost all afternoon today so I will be able to tell tomorrow if I have solved the problem.

I got my mast raised and the stays measured so I could order the cable for them. As soon as that comes in I can finish with the mast. Then I think I will hook all my wiring and antenna leads up and check to see that everything is working. I will have to take my computer over to the boat and install it there to check everything out. In the meantime, Linda and I have been loading our belongings into the boat and getting it ready to move. I still don't have a firm date when the movers will pick up the boat; it's still either the last week of May or the first week of June. I called the movers today but they haven't called me back yet. There is only one week left until the last week of May and I would really like a little better idea of when I need to have everything ready to take apart for the move.

Monday May 26, 2014

I got my mast stays finished and I raised the mast with the forestays and aft stays installed. I temporarily installed the boom just to see if I made it long enough to reach over the side so I could lift my dingy without it hitting the side of the boat. It looks like I guessed about right for the length of the boom. I took a couple pictures of the mast which are on the next page.

Here are the two pictures I mentioned. You can just make out the stays holding the mast in place.

I got the word from the movers and they said they would be here on the 4th of June at 8:00 in the morning to pick up the Molly B. Now I know when I really need to have this boat ready for transport.

Wednesday & Thursday, June 4th & 5th, 2014

This was supposed to be the day the boat got moved to Ft. Pierce on the coast. The moving company decided they didn't have a truck in the area so they cancelled the move until they could get a truck. They had already scheduled the crane for today, and they called them and canceled the crane but they never bothered to call me and let me know that the date was going to be changed. I was livid and got them on the phone after waiting half the morning for them to show up. They explained to me what had happened and I told them this was not acceptable because I had people flying in from up north and I couldn't ask them to change their flights. I told the movers that my boat would move now as they had promised and I didn't care if they had to rent a truck to do it. I also called the crane company and they agreed to try to get a crane over to Frostproof if the movers could get a truck there. After a long day on the phone back and forth, they finally said they had a truck that they could have there for the move to take place on the 5th. I agreed and the crane company said they would have a crane there by 7:30 in the morning. Then I called Cracker Boy Boat Works in Ft. Pierce to arrange for a crane and the travel lift to unload the truck

with and they said they couldn't do it on the 5th because they were booked solid all afternoon with the travel lift. I tried to find another yard I could use to no avail and then I tried to hire another crane company over there to unload the truck with and that didn't work so I called Cracker Boy back and they said if the driver could have the boat there my 11:00am they would be able to get it unloaded. The big rush to unload was the fact that the driver had to be in Wisconsin by Saturday and on the Canadian border by Sunday morning to take another boat into Canada. The driver agreed to try to make it by the deadline. To the left is a picture of the truck with a 53 foot trailer behind waiting for the crane to show up early on the 5th. To the left on the next page is a picture of the crane trying to get into position to lift the boat. As you can see he didn't get very far before he sank into the sugar sand and got really stuck.

The picture below is a little closer view of the stuck crane. They jacked it up with the outriggers and put some planking under the wheels and then pulled it out with another truck. The second attempt resulted in the same predicament and they had to pull him out again. They tried a third time and while they didn't get it into the position they wanted, they go it close enough to do the job. All this took most of the morning so there was no way the driver was going to make the 11:00 am deadline.

Once they got the crane into position and set up, the first lift they made was the

aft section of the fly bridge. The picture to the left shows it coming off the boat. They set the fly bridge section aside and moved the truck that was going to haul it into position and lifted the top of the raised pilot house. They needed the truck and trailer in position because they were going to set the main bridge roof on a platform atop the fifth wheel of the truck. To the right is a view of the bridge roof being loaded onto the platform over the fifth wheel. Next was the main lift of the boat. This took them about an hour just to rig the sling to lift the boat with. More time was going by than work being done. The truck driver said they were really doing great, and that most crane outfits would take two days or more to do the same lift. So, although it seemed slow to me, apparently they were moving right along. As you can see in the pictures on the next page, we had quite a crowd watching the Molly B being loaded.

These are a few of our very good friends that came by to watch the loading of the Molly B onto the truck, headed for the coast. It was quite a project to watch even though it took all of the morning and into the afternoon before we were on the road with the whole boat on the truck.

Here are a couple pictures of the slings and spreader bars they rigged up to lift the boat with. They had to raise the boom on the crane quite a ways to get all the slack out of the slings. The picture below shows how high they raised the boom.

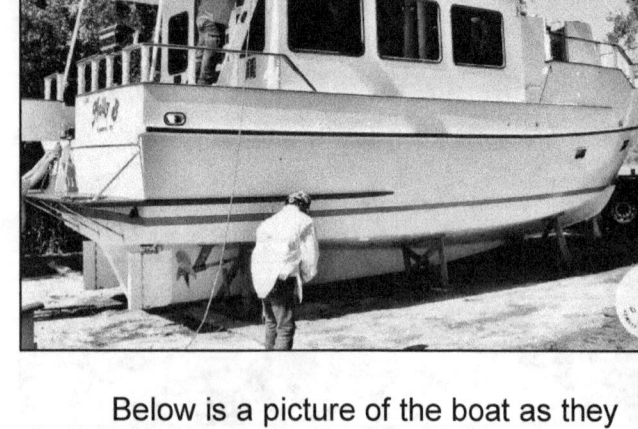

Below is a picture of the boat as they started the lift. This was the part that worried me the most, but everyone on the crane crew seemed to know what they were doing and it went really well.

Below is a sequence of pictures showing the loading process. They had to swing the boat around so they could load it stern first on the trailer. Then they loaded the aft section of the fly bridge on the back end of the trailer under the bow of the boat.

The Boat is off the cribbing

Turning the boat around

Boat is ready to load Stern first

Coming over the trailer   *Almost lined up and loaded*   *She's Loaded, adjusting pads*

To the left the last piece is being loaded. This is the aft section of the fly bridge. We are almost ready to hit the road.

We were a little late getting started after the truck left with the boat, but we caught up to him just a little north of Frostproof on county road 630. Following behind the truck, that boat really looked like it took up the whole road. Several big rigs actually pulled partially off the road to let the boat pass. To the right is a view of the load going down the road. Looks pretty big but the driver managed to make good time at around 55 mph.

We have arrived at Cracker Boy Boat Works in Ft. Pierce, Florida in the picture to the left and we are waiting to see if we can get unloaded yet today or if we will have to wait until tomorrow. We were way beyond out 11:00 o'clock deadline but we were hoping we could get unloaded before the yard closed at 5:00 o'clock. They had to finish spotting a 72 foot yacht before they could unload us.

To the left is a view of the truck and my boat under the travel lift ready to be unloaded. They finished spotting a 72 Ft. yacht a little ahead of schedule so they were able to get us unloaded.

While we were waiting, they unloaded the bridge roof and the aft section of the fly bridge so once we got the word, all they had left to do was unload the boat from the trailer and the driver could head out. To the right is a view of the boat coming off the trailer. Now the driver can pull out and hit the road for his next pickup.

To the left is a view of the boat moving down the road to where they are going to spot it on the hardstand while I put it all back together again. It's now right at 5:00 o'clock so they didn't get it spotted before they closed the yard. That meant we couldn't stay aboard the boat for the night so we had to find a motel.

Friday, June 6, 2014

To the right is a picture of my brother Steve watching while they spot the boat. Once they had it spotted we started working on getting it put back together and the painter Bill started on the bottom paint. I'm having the yard do the bottom paint for me so I can concentrate on getting the boat put back together. Steve flew down from Minnesota today and Linda went up to Orlando and picked him up at the airport. It sure is great he's here, I really need the help!

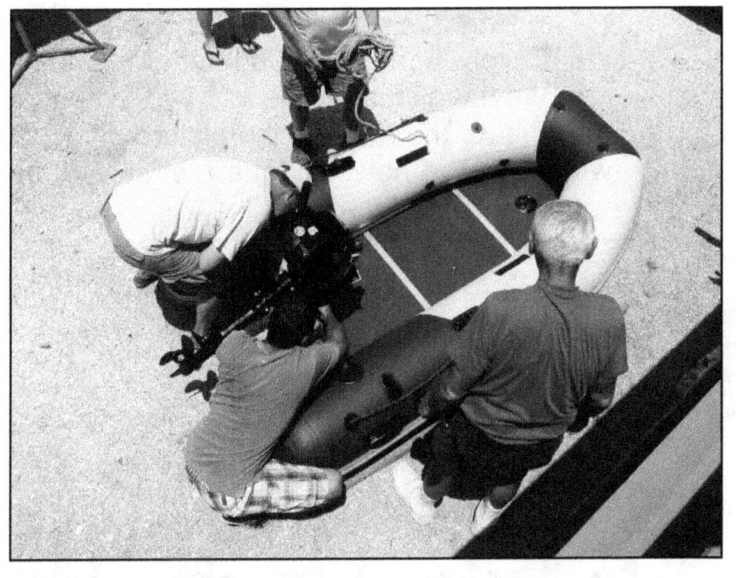

To the left is a picture of everybody working on the dingy getting it ready to put up on the boat deck once we have the mast up and the winches rigged. My good friend Woody and his wife Pat drove all the way down here from Michigan to see the Molly B launch. He drove his truck down so he hauled my dingy, motor, and doors off the boat over to Ft. Pierce for me. We got his truck unloaded and decided to mount the motor and get it ready. My oldest boy Tim and my youngest boy Jay are working on mounting the motor and Steve and I are supervising.

Monday June 9, 2014

Everybody has really been busy over the weekend helping me put the Molly B back together. Jason's wife Tracey and our two grandkids Paton and Jessica had a good time watching us work and they even made it to the beach a couple times. Tim and Jason had to leave to get back to work so now it's just Steve and I left to finish up. To the right is a view of Steve and I on the aft deck after almost everything is put back together. We are almost ready for the launch. Bill, the painter has the primer all put on the bottom and just has to put the last coat of paint on and we will be ready.

Wednesday June 11, 2014

Below is a picture of the Molly B all ready for the launch. Everything is back together and the bottom paint is finished. I have the launch scheduled for 2:00 o'clock this afternoon. It turned out that the yard had two coast guard boats that they needed to haul out so they asked me if I wanted to move the launch up to 1:00 o'clock and I said sure so they got the travel lift and loaded her up. We have some friends coming over from Frostproof to watch the launch and moving it up an hour might make them miss the big event. I hope they get here a little early.

Here we go, we are all loaded onto the travel lift and headed for the launch ramp in the picture to the left. We're right at the front of the ramp in the picture below. It won't be long now, and the Molly B will be wet for the first time!

We're over the water in the picture to the left and now Steve and I have to get onboard so we can make sure we don't have any leaks before they pull the straps out from under the boat.

The Molly B is launched in the picture below and she's floating just about where I thought she would. I could have

raised the water line just a bit but otherwise she is floating as designed, and with only one minor leak on the starboard shaft seal. Hopefully that is a seal problem that will go away once we get underway. In the final picture of the launch to the left above, I'm backing out of the launch ramp.

The launch went great although it wasn't without some problems. I had the transmission control cables hooked up backwards so when I had the control levers in the reverse position, I went forward and in the forward position I went back. That was just a little confusing to say the least, trying to dock alongside the pier at the launch ramp. I finally got tied up at the pier and started working on the control cables. To the left is a picture of me pulling up to the pier after backing out of the launch ramp. Below is a picture of us moving to the marina across the creek to take on some fuel before heading north to Vero Beach Marina. Below is a view of us heading out the channel for the ICW heading north. This was an unbelievable moment after 11 ½ years of work, I'm finally on the water and underway with a boat I built myself.

Below is a picture of the Molly B docked at the Vero Beach City Marina. It was a pleasant trip up to Vero Beach with

Steve doing most of the driving and me running around the boat checking everything out while underway. The first trip on the water wasn't without some problems as well. The boat wasn't answering the helm very well, and steerage was difficult to say the least. The farther along we got, the worse the steering became. I concluded we must have a hydraulic leak somewhere. When I made the final turn into the channel to the marina, I had lost almost all steerage. When I got to the slip I had to dock using the engines because my steering was no longer working. As you can see in the picture above the mast is down because I had to move the aft section of the fly bridge back so I could access the hydraulic lines under the deck. I found the leak in the quick disconnects under there so I removed them and replaced them with couplers because I don't think I'm going to ever have to remove the bridge roof again. Once I refilled the system with hydraulic fluid the steering worked great. I also had a pretty good leak at the shaft seal on the starboard engine. I installed a spare seal carrier on the shaft when I installed them so all I had to do was remove the old seal and slide the new one into the holder. I wasn't sure I wanted to attempt that maneuver myself with the boat in the water, so I called a guy who said he had done it before and would be happy to do it for me. When he arrived he discovered that "whoever installed" the seal the first time had forgotten to remove the installation color around the seal. Once that was removed, the leak was fixed. I don't know where I was the day I did that, but I must have been somewhere else. Once that was fixed, I pumped out the bilges and I thought I was all set. Then I discovered my VHF radio wasn't working very well, and a radio guy determined it was my receiver that was bad so I had to replace the radio. It's working great now. I guess this is why you do sea trials before you head out on your first voyage. I've also found some other minor problems with my computer, AIS system, and my generator. We've been here at Vero for just over two weeks now fixing things and doing sea trials. I'm waiting for parts for my generator right now, the 28th of June, and as soon as they arrive and I get my generator working we will be heading north. We can't wait to get underway.

This marks the end of the Building Journal of the Molly B and the beginning of a new chapter in our lives aboard our new home. Linda and I will attempt to keep you up to date on our adventures on our blog at *http://buildingthemollyb.blogspot.com/*.

Fair Winds and Following Seas to all.

Source List

Trying to find the materials to build your boat can sometimes be a problem so I have included a list of suppliers that I have used in the construction of the Molly B to date.

Boat Plans & Misc. Supplies

Glen L Marine Designs
9152 Rosecrans Ave.
Bellflower, CA 90706
(562)630-6258
www.glen-l.com

Epoxy & Paint

Merton's Fiberglass Supply
PO Box 399
East Longmeadow, MA 01028
(413)736-0348
www.mertons.com

Fasteners

McFeely's Square Drive Screws
3720 Cohen Pl
Lynchburg, VA 24501
(800)443-7937
www.mcfeelys.com

Manasquan Premium Fasteners, LLC
PO Box 669
Allenwood, NJ 08720
(800)542-1979
www.manasqualfasteners.com

CC Fasteners
16 Aqua Lane
Tonawanda, NY 14150
(800)992-5151
www.ccfast.com

Sanding Discs & Tools

Grizzly Industrial, Inc.
PO Box 2069
Bellingham, WA 98227
(570)546-9663
www.grizzly.com

Boat Hardware, Tools & Misc. Supplies

Jamestown Distributors
500 Wood Street, Bldg. 15
Bristol, RI 02809
(401)253-3840
www.jamestowndistributors.com

Plastic Rod

Modern Plastics, Inc.
Corporate Headquarters
678 Howard Ave.
Bridgeport, CT 06605
(800)234-9696
www.modernplastics.com

Xynole Polyester Flexible Fabric

Defender Marine Outfitter
42 Great Neck Road
Waterford, CT 06385-3336
(800)628-8225
www.defender.com

Marine Plywood & Lumber

World Panel Products, Inc.
1750, #1 Australian Ave.
Riviera Beach, FL 33404
(888)863-3379
www.worldpanel.com

Noah's Marine
54 Six Point Rd.
Toronto, ON M8Z2X2
(416)232-0522
www.noahsmarine.com

Logan Lumber Co.
PO Box 1608
Tampa, FL 33601
(813)253-3445

Reference List

Below are just a few of the reference books I have used to acquire the knowledge and skills needed to build a boat. You will find a hundred ways of doing everything connected with the construction of a boat in these books. Put them all together and pick what works best for you. Boatbuilding is not as mysterious as you may think.

The New Cold-Molded Boatbuilding
Reuel B. Parker

The McGraw-Hill Companies
Customer Service Department
P.O. Box 547
Blacklick, OH 43004
1-800-262-4729

Thirty-four chapters discuss the New Cold-Molded Boatbuilding method methodically and in depth, and cover everything from choosing a design to rigging, wiring, canvas work, and launching. Nothing is left out. Appendices show how to plan a project, where to find materials and supplies, and include a section of the authors designs.

Understanding Boat Design
Ted Brewer

The McGraw-Hill Companies
Customer Service Department
P.O. Box 547
Blacklick, OH 43004
1-800-262-4729

"One of the cleanest and clearest expositions on the elements of yacht design ever published ... by a naval architect who knows what he is talking about"
---- Wooden Boat

The elements of Boat Strength
Dave Gerr

The McGraw-Hill Companies
Customer Service Department
P.O. Box 547
Blacklick, OH 43004
1-800-262-4729

Acclaimed author and naval architect Dave Gerr created this unique system of easy-to-use scantling rules and rules-of-thumb for calculating the necessary dimensions, or scantlings, of hulls, decks, and other boat parts ... The elements of boat strength offers their context: in-depth, plain-English discussion of boatbuilding materials, methods, and practices that will guide you through all aspects of boat construction.

Buehler's Backyard Boatbuilding
George Buehler

The McGraw-Hill Companies
Customer Service Department
P.O. box 574
Blacklick, OH 43004
1-800-262-4729

If you want to build a simple, rugged, economical, good-looking cruising boat---power or sail---using every day lumberyard materials and few skills other than perseverance, this is the book for you.

Boatbuilding With Plywood
Glen L. Witt

Glen L. Marine Designs
9152 Rosecrans
Bellflower, CA 90706

Most texts on boatbuilding discuss plywood and its application to boatbuilding in a short paragraph, or at best in a single chapter. This book, however, will show in detail the "how-to" of boat construction with the emphasis on the use of plywood.

How to fiberglass Boats
Ken Hankinson

Glen L. Marine Designs
9152 Rosecrans
Bellflower, CA 90706

The subject of this book is how to cover or sheath boats with fiberglass when used in conjunction with various types of resins. Yet these materials have many other uses. Thus, the information presented may be considered as a handbook to using these materials in a variety of applications, with perhaps some minor modifications on the part of the user to suit his specific requirements.

Propeller Handbook
Dave Gerr

The McGraw-Hill Companies
Customer Service Department
P.O. Box 547
Blacklick, OH 43004
1-800-262-4729

"This book is for everyone who has ever had to make a decision about a propeller; Mechanics, boat builders, boat service yard owners, boat owners, as well as naval architects. Dave Gerr and International Marine made a complicated topic understandable and put it into a handbook that is easy to use' ---Wooden Boat

Inboard Motor Installations
Glen L. Witt / Ken Hankinson

Glen L Marine Designs
9152 Rosecrans
Bellflower, CA 90706

The inboard motor, together with its power transmission system, related components, and installation requirements is a mystery to many boat owners, amateur builders, and even some boating professionals. The many types of power plants and transmission choices available seem to complicate the picture for most novices. Yet like so many things in today's complex world, when taken a step at a time, the installation of an inboard motor in a boat is relatively simple

www.ingramcontent.com/pod-product-compliance
Lightning Source LLC
Chambersburg PA
CBHW080942170526

45158CB00008B/2348